the mad scientist handbook

D0030209

ALSO BY JOEY GREEN

Hellbent on Insanity

The Unofficial Gilligan's Island Handbook

The Get Smart Handbook

The Partridge Family Album

Polish You Furniture with Panty Hose

Hi Bob!

Selling Out: If Famous Authors Wrote Advertising

Paint Your House with Powdered Milk

Wash You Hair with Whipped Cream

The Bubble Wrap Book

Joey Green's Encyclopedia of Offbeat Uses for Brand-Name Products

The Zen of Oz: Ten Spiritual Lessons from Over the Rainbow

The Warning Label Book

Monica Speaks

The Official Slinky Book

You Know You've Reached Middle Age If . . .

the mad scientist Handbook

How to Make Your
Own Rock Candy,
Antigravity Machine,
Edible Glass, Rubber
Eggs, Fake Blood,
Green Slime, and Much,
Much More

joey green

A PERIGEE BOOK

Most Perigee Books are available at special quantity discounts for bulk purchases for sales promotions, premiums, fund-raising, or educational use. Special books or book excerpts can also be created to fit specific needs.

For details, write to Special Markets, The Berkley Publishing Group, 375 Hudson Street, New York, New York 10014.

WARNING: A responsible adult should supervise any young reader who conducts the experiments in this book to avoid potential dangers and injuries. The author has conducted every experiment in this book and has made every reasonable effort to ensure that the experiments are safe when conducted as instructed; however, neither the author nor the publisher assume any liability for damage caused or injury sustained from conducting the projects in this book.

A Perigee Book
Published by The Berkley Publishing Group
A division of Penguin Putnam Inc.
375 Hudson Street
New York, New York 10014

Copyright © 2000 by Joey Green
Photographs by Joey Green,
except photograph on page xi by Debbie Green
Book design by Richard Oriolo
Cover design by Steve Ferlauto
Cover art by Brad Weinman

First edition: April 2000

Published simultaneously in Canada.

The Penguin Putnam Inc. World Wide Web site address is
http://www.penguinputnam.com

Library of Congress Cataloging-in-Publication Data

Green, Joey.
    The mad scientist handbook : How to make your own rock candy, antigravity machine, edible glass, rubber eggs, fake blood, green slime, and much, much more / Joey Green.—1st ed.
    p.   cm.
    Includes bibliographical references.
    ISBN 0-399-52593-9
    1. Science—Experiments. I. Title.

Q164.G73 2000
793.8—dc21   99-057902
                                                                          CIP

Printed in the United States of America

10 9 8 7 6

For

Ashley, Julia,

Eric, Rebecca, Andrew, Lauren,

Matthew, Sammy,

Jonathan, Alexander, Zachary,

Corrine, Janelle,

Kailey, and Shannon

the mad scientist handbook

contents

introduction

When I was in the fourth grade, I nearly blew up our house. My father helped me build a model volcano with metal screening in the shape of a cone tacked to a piece of wood, strips of newspaper, plaster of Paris, and brown paint. He bought a jar of strange orange powder from a chemical supply store. I poured the powder into the upside-down metal cap from an aerosol spray can built into the top of the volcano, lit it with a match, and *KA-BOOM!*

While I didn't learn all that much about real volcanoes, I did learn that my dad was one very cool guy, although he never told me where on earth he bought that strange orange powder or what it was called—until now.

This book is for anyone who loved all those pointless, wacky mad science experiments we did as kids. It's a simple one-stop cookbook for making fake blood, green slime, edible glass, quicksand, underwater fireworks, and a huge mess in your kitchen and garage—all with tools and brand-name products you've already got around the house (or stuff you can easily get from a chemical supply house listed in your local Yellow Pages). I've tested everything in this book. It all works. Guaranteed.

If you happen to learn some weird scientific principles along the way, don't blame me. Sure, I've provided short, geeky explanations of why everything in this book works, in case you just have to know. But all I'm really interested in teaching you is strange ways to have fun.

—JOEY GREEN

antigravity machine

WHAT YOU NEED

Three books, each at least 1 inch thick

Two yardsticks

Two plastic funnels of equal size

Black electrical tape

WHAT TO DO

Stack two books on top of each other on the floor. Place the third book far enough away on the floor so you can lay a yardstick across the books to form a bridge. Place the second yardstick next to the first yardstick to form a V-shape with the open end of the V on the stack of two books. Tape the bowls of the funnels together. Place the joined funnels on the lower end of the track formed by the yardsticks.

WHAT HAPPENS

The joined funnels roll up the incline.

WHY IT WORKS

Although the joined funnels appear to defy the laws of gravity, in reality, their center of gravity (the point at which the effect of gravity on an object seems to be concentrated) moves downward as the joined funnels move along the inclined yardsticks.

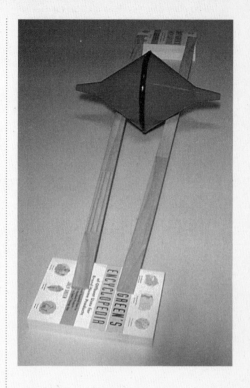

BIZARRE FACTS

■ The center of gravity of the hollow joined funnels is at its center, even though there is no matter at that point for gravity to affect.

■ When a boomerang is thrown, it spins about its center of gravity, which is outside its body, between the arms of the V.

■ The Moon's gravitation causes the ocean tides on Earth.

■ You would weigh 3.75 percent more if you were standing on the North Pole than you would if you were standing on the equator. An object on earth does not weigh the same at all places on the planet because the earth rotates and it is not perfectly round.

- While Galileo Galilei is credited with determining that falling objects fall at the same rate, Giambattista Benedetti determined the exact same thing in 1553—eleven years before Galileo was born.

- Sir Isaac Newton, considered a poor student in school, discovered gravity, invented calculus, and, in 1699, became master of the mint in England, prosecuting counterfeiters.

THE FALL OF ARISTOTLE

The Greek philosopher and scientist Aristotle claimed that heavy objects fall faster than light objects, a widely accepted belief until the sixteenth century when, according to legend, Italian scientist Galileo Galilei simultaneously dropped a hammer and a feather from the Leaning Tower of Pisa (supposedly proving that all objects fall at the same rate of acceleration).

bALLooN caR

WHAT YOU NEED

Krazy Glue

Small plastic spool

Old compact disc

Button

Medium-sized toy balloon

WHAT YOU DO

Using the Krazy Glue, glue the plastic spool onto the compact disc so the hole in the spool is directly over the hole in the compact disc. Let dry. Glue a button over the top hole in the spool so the holes in the button are directly over the hole in the spool. Inflate the balloon, pinch the neck to prevent the air from escaping, and stretch the lip of the balloon over the spool. Set the compact disc on a flat tabletop and let go of the balloon.

WHAT HAPPENS

The compact disc floats across the table like a hovercraft.

WHY IT WORKS

An invisible cushion of air acts as a lubricant and reduces friction between the compact disc and the tabletop, the same way adding oil to a car engine prevents the parts from rubbing against each other.

BIZARRE FACTS

■ The word *balloon* is slang for "a hobo's bedroll."

■ A compact disc holds three miles of playing track and is read from the inside edge to the outside edge, the reverse of how a record works.

■ In 1954, English electronics engineer Christopher Cockerell designed the hovercraft by attaching two tin cans, one inside the other, to an industrial air blower mounted on a stand and blowing air through the gap between the tin cans.

■ The SRN-4 Mark III, the world's largest civil hovercraft, weighs 305 tons, carries more than four hundred passengers and sixty cars across the English Channel, and travels at a top speed of seventy-five miles per hour—nearly twice the speed of the fastest oceanliner.

bALLooN iN A bottle

WHAT YOU NEED

Clean, empty 2-liter soda bottle

Electric drill with $1/8$-inch bit

Medium-sized toy balloon

WHAT YOU DO

With adult supervision, drill a hole in the center of the bottom of the soda bottle. Insert the balloon into the bottle, folding the neck of the balloon over the bottle's mouth. Inflate the balloon inside the bottle by blowing into it. Immediately place your thumb over the hole in the bottom of the bottle and remove your mouth from the mouth of the bottle. Then remove your thumb from the hole.

WHAT HAPPENS

While your thumb is over the hole in the bottom of the bottle, the balloon remains inflated inside the bottle with its mouth open. When you remove your thumb, the balloon instantly turns inside out to inflate again above the bottle.

WHY IT WORKS

The hole in the bottle lets the air escape as you inflate the balloon.

Placing your thumb over the hole prevents air from entering the bottle to deflate the balloon.

BIZARRE FACTS

- If you fill the inflated balloon with water and then remove your finger from the hole in the bottom of the bottle, the increased air pressure will push the water out.

- Another variation: With adult supervision, use a funnel and wear an oven mitt to carefully pour a cup of boiling water into a clean, empty, large, glass narrow-necked bottle. Swirl the water inside the bottle, then pour the water out of the bottle. Immediately attach the neck of a balloon over the bottle's mouth. The air inside the bottle contracts while it cools, drawing the balloon inside, turning it inside out, and inflating it.

- Air pressure changes the boiling point of water. At sea level, the boiling point of water is 212°, but as the height above sea level increases, the temperature required to boil water also increases, making it dif-

ficult to bring water to a boil at high altitudes.

- Robert Boyle, the seventeenth-century Irish scientist who formulated Boyle's law regarding air pressure, referred to atoms as "corpuscles."

- Model ships in narrow-necked bottles are made by inserting the finished model into the bottle with its hinged masts and sails lying down, then pulling a thread to draw the rigging upright.

- Laidley, Australia, is home to the largest bottle in the world. Measuring 6 feet 11 inches tall and 5 feet $4\frac{1}{2}$ inches in circumference, the bottle holds ninety-two gallons of liquid.

PEAR IN A BOTTLE

Eau de Vie de Poire, a narrow-necked bottle of French liqueur containing a whole pear, is made by placing the bottle over the budding fruit while it is still growing on the tree, suspending the bottle in a net by tying it to a branch above, waiting for the pear to mature, and then filling the bottle with liqueur.

bALLooN rocket

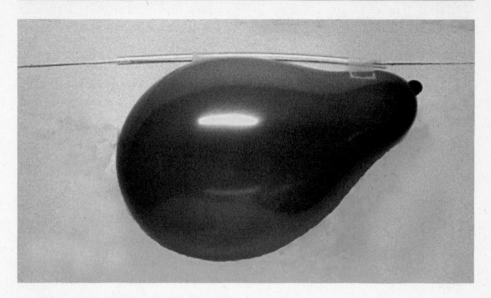

WHAT YOU NEED

Balloon

Piece of string, 10 to 25 feet long

Drinking straw

Scotch tape

WHAT TO DO

Tie one end of the string to a tree or post. Thread the straw onto the free end of the string, and then tie that end of the string to a second tree or post, making sure the string is taut. Move the straw to one end of the string.

Inflate the balloon and pinch the neck to prevent the air from escaping. Use two pieces of Scotch tape to attach the inflated balloon to the straw so the balloon is parallel to the straw and the mouth of the balloon points toward the closest post or tree.

Release the balloon.

WHAT HAPPENS

The balloon and straw jet across the string until the balloon completely deflates.

WHY IT WORKS

As Newton's third law states, "For every action there is an opposite and equal reaction." When you release the neck of the balloon and the compressed air rushes out into space, the reaction to it drives the balloon forward.

BIZARRE FACTS

- The basic principle behind a balloon zooming across a string is exactly the same principle behind a space rocket launching into space. When the fuel burns, gas escapes from the rocket's bottom, pushing the rocket upward.

- The cruise liner *Queen Elizabeth II* moves only six inches for each gallon of diesel fuel it burns.

- The longest recorded flight of a chicken is thirteen seconds.

- In 1895, Lord Kelvin, president of the Royal Society, said: "Heavier-than-air flying machines are impossible."

- In 1921, responding to rocket scientist Robert Goddard's revolutionary work, the *New York Times* editorialized: "Professor Goddard does not know the relation between action and reaction and the need to have something better than a vacuum against which to react. He seems to lack the basic knowledge ladled out daily in high schools." Five years later, Goddard launched the first liquid fuel rocket.

- Felix the Cat was the first cartoon character ever to be made into a balloon for a parade.

IN SPACE NO ONE CAN HEAR YOU SCRIBBLE

During the space race in the 1960s, NASA spent $1 million to develop a ballpoint pen that would write in zero gravity. The Soviet Union solved the same problem by giving their cosmonauts pencils.

copper NAiL

WHAT YOU NEED

Clean, empty glass jar

¼ cup white vinegar

⅛ teaspoon salt

Twenty copper pennies

Iron nail

Baking soda

Sponge

WHAT TO DO

Pour the vinegar and salt into the jar. Stir well. Put the pennies into the vinegar-and-salt solution for three minutes. Clean the nail with the baking soda and sponge. Rinse thoroughly. Drop the clean nail into the vinegar-and-salt solution with the pennies. Wait fifteen minutes.

WHAT HAPPENS

The nail is coated with copper, and the pennies are shiny clean.

WHY IT WORKS

The acetate in the vinegar (also known as acetic acid) combines with the copper on the pennies to form copper acetate, which then adheres to the iron nail.

BIZARRE FACTS

- The word *vinegar* is derived from the French words *vin* (wine) and *aigre* (sour).

- The oldest way to make vinegar is to leave wine made from fruit juice in an open container, allowing microorganisms in the air to convert the ethyl alcohol to acetic acid.

- Vinegar lasts indefinitely in the pantry without refrigeration.

- Hannibal, the Carthaginian general, used vinegar to help clear boulders blocking the path of his elephants across the Alps. Titus Livius reported in *The History of Rome* that Hannibal's soldiers heated the rocks and applied vinegar to split them.

- According to the New Testament, Roman soldiers offered a sponge filled with vinegar to Jesus on the cross. While the act is considered cruel, vinegar actually shuts off the tastebuds, temporarily quenching thirst, suggesting that the Roman soldiers may have been acting out of kindness.

- Copper is the best low-cost conductor of electricity.

- Copper does not rust. In damp air, copper gets coated with a green film called a *patina*, which protects the metal against further corrosion.

- Pennies are actually made from zinc coated with less than 3 percent copper.

- Soaking pennies in Taco Bell hot sauce removes the tarnish.

....

HEADS, YOU LOSE

A penny will be tossed heads 49.5 percent of the time. The head side weighs 0.5 percent more than the tail side, so it tends to land downward.

coʃmic ʀAY detector

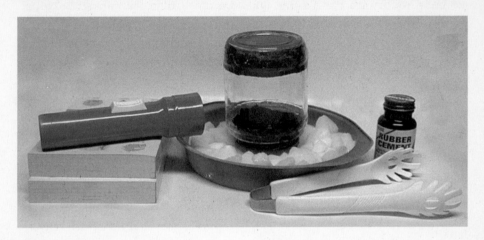

WHAT YOU NEED

Scissors

Black heavy felt

Clean, empty glass jar with a rubber washer or cardboard filler under the lid

Ruler

Rubber cement

Black velvet

Rubbing alcohol

Two towels

Pie tin

Rubber gloves

Block of dry ice (enough to fill a pie tin)

Hammer

Ice tongs

Flashlight

Three books

Masking tape

Small magnet

WHAT TO DO

Cut the felt in a circle to lie in the bottom of the jar. Glue the felt in place with rubber cement. Cut a strip of felt 1 inch wide and glue it around the inside wall at the bottom of the jar.

Cut a strip of velvet 1 inch wide and glue it to the top of the jar's inside wall. Cut a circular piece of velvet to fit inside the metal lid, over the rubber or cardboard, and glue it in place.

Pour enough rubbing alcohol into the jar to saturate the felt at the bottom of the jar thoroughly and cover the bottom of the jar. Screw the lid on the jar tightly and let it sit for ten minutes.

Spread a towel on a flat surface and place the pie tin on top of it. Wearing rubber gloves (to avoid touching the dry ice with your bare hands), wrap the second towel around the block of dry ice, then break up the block with the hammer. Using the ice tongs, put the cubes of dry ice into the pie tin, making a level surface on which to place the jar.

Turn the jar upside down and place it on the dry ice in the center of the pie tin. Position the flashlight on top of the books, attaching it in place with masking tape, aiming the beam through the lower half of the jar. Turn off the lights, and cover the bottom of the jar with the palm of your hand to warm the alcohol-soaked velvet. Watch carefully.

WHAT HAPPENS

Within five minutes, the alcohol vapor condenses, warmed by your hand, and you'll see a continuous rain of fine mist 1 inch below the top of the chamber. After another five minutes, the rain decreases.

When the interior temperature is just right, you see cobweblike threads suddenly appearing and disappearing at various angles about an inch above the lid. These are vapor trails made by cosmic rays passing through the jar. Place the magnet against the side of the jar, and the trails will be deflected toward it.

WHY IT WORKS

The heat from your hand warms the top of the chamber, and the dry ice cools the bottom of the chamber. Somewhere between these tem-

perature extremes, usually about 1 to 2 inches from the bottom, the air becomes saturated with alcohol vapor, and particle trails become visible where cosmic rays cause the alcohol to condense.

BIZARRE FACTS

- Cosmic rays are high-energy particles that originate from explosions in outer space—from phenomena such as supernovas and pulsars. At this very moment, cosmic rays are hurtling through space at nearly the speed of light.

- Trillions of these particles pass through the earth's atmosphere every few minutes.

- Three to six cosmic ray particles strike each square inch of the earth's atmosphere every second.

- Cosmic rays are penetrating your body at this very moment.

- Physicist Murray Gell-Mann named the subatomic particles known as quarks after a line from James Joyce's novel *Finnegans Wake*: "Three quarks for Muster Mark!"

- The common goldfish is the only known animal that can see both infrared and ultraviolet light.

- Adolf Hitler, refusing to accept the fact that a Jew had come up with the theory of relativity, falsely claimed that Albert Einstein had stolen the idea from some papers found on the body of a German officer who had been killed in World War I.

COSMIC GILLIGAN

When a meteorite hits Gilligan's Island, the Professor, after measuring the meteorite with a bamboo Geiger counter, claims: "There were cosmic rays, which aren't as deadly as interstellar radiation; however, they can kill you." The Professor is just plain wrong. Cosmic rays are a form of radiation that is far too weak to endanger anyone.

edible gLaSS

WHAT YOU NEED

Butter

Baking sheet

1 cup sugar

Heavy stainless steel or nonstick frying pan

Large wooden spoon

WHAT TO DO

Butter the baking sheet, and place it in the refrigerator.

Put the sugar in the frying pan. With adult supervision, set the pan on a burner at low heat. Stir the sugar slowly as it heats up. The sugar will slowly turn tan, stick together in clumps, and begin melting into a pale brown liquid. Continue stirring until the sugar melts into a thick brown liquid.

Pour the brown liquid into the cold baking sheet. Let cool.

WHAT HAPPENS

The melted sugar hardens into a sheet of edible sugar glass.

WHY IT WORKS

Sugar is made of crystals, just like glass, which is made from sand.

BIZARRE FACTS

- Most glass is made from a mixture of silicon dioxide (the main ingredient in sand), soda (sodium oxide), and lime (calcium oxide).

- Silicon dioxide is one of the most inexpensive, most plentiful materials on earth.

- Fiber optic cable is made from glass and carries far more information than the same size wire cable.

- In Hollywood films, fake glass windows and fake glass bottles broken over the heads of movie stars were originally made from sugar. Today, they are made from a special resin.

- Glass can be made more fragile than paper or stronger than steel.

- Butter was probably discovered by accident. When milk is transported in containers, the agitation naturally makes the cream congeal.

MR. EAT-IT-ALL

Gastroenterologists have confirmed that Michael Lotito of Grenoble, France, has the uncanny ability to eat and digest glass and metal. Since 1966, Lotito has eaten seven television sets, six chandeliers, a computer, ten bicycles, a supermarket cart, a Cessna aircraft, and a coffin. He has been nicknamed Monsieur Mangetout—French for Mr. Eat-It-All.

egg in a bottle

WHAT YOU NEED
Water

Teakettle

Oven mitt

Funnel

Clean, empty salad dressing bottle

Hard-boiled egg, peeled

WHAT TO DO

With adult supervision, boil water in the teakettle. Wearing the oven mitt and using the funnel, carefully fill the empty salad dressing bottle with the boiling water from the teakettle. Remove the funnel. Swirl the water around inside the bottle, then pour the water into the sink. Quickly place the egg over the mouth of the bottle.

WHAT HAPPENS

The egg is sucked into the bottle, making a very unusual sound. (To get the egg out of the bottle, hold the bottle upside down and blow into the bottle for thirty seconds. When you remove your mouth, the increased air pressure in the bottle forces the egg out of the bottle.)

WHY IT WORKS

The heat from the boiling water causes the air inside the bottle to expand, forcing some of it out. As the air begins to cool inside the bottle, it contracts, reducing the air pressure inside the bottle. The greater air pressure outside the bottle forces the egg into the bottle.

BIZARRE FACTS

■ The tradition of exchanging colored eggs in the springtime predates Easter by several centuries.

■ The ancient Egyptians buried eggs, a symbol of resurrection and birth, in their tombs.

■ The ancient Greeks placed eggs atop graves. When the Greeks took over ancient Israel, many Jews adopted Hellenistic practices. To this day, Jews place rocks atop gravestones to signal that the grave has been visited and the loved one remembered (perhaps as a substitute for eggs).

- Legend holds that Simon of Cyrene, the egg merchant who helped carry Christ's crucifix to Calvary, returned to his farm to discover that all of his hens' eggs had turned to a rainbow of colors.

- Raw eggs harden when boiled in water because the water's intense heat causes the egg's protein strands to unravel, exposing their ends, which then bond together with other unraveled protein strands.

- Since the protein structure of egg white and egg yolk vary slightly, the egg white hardens at 176°F, while the egg yolk hardens at 185°F.

- According to *The Guiness Book of Records*, White Horse Scotch Whiskey makes the smallest bottles of liquor now sold in the world. The bottles are two inches high and contain 1.184 milliliters (less than one-eighth teaspoon).

HOAX IN A BOTTLE

A note found in a bottle on the coast of Denmark in 1946, written on a page torn from the logbook of the German U-boat *Naueclus* and dated one year earlier, claimed that Adolf Hitler did not die in the Berlin bunker but aboard the *Naueclus*, which sank on November 15, 1945, while sailing from Finland to Spain.

electric Lemon

WHAT YOU NEED

Wire cutters

Five stiff copper wires, 6 inches long, 14 grade

Four galvanized nails

Four lemons

Bulb holder

1.2-volt flashlight bulb

Repeat until all the lemons are wired together. Attach the remaining two wires to the bulb holder, and insert the bulb.

WHAT TO DO

With the wire cutters, strip 1 inch of insulation off both ends of each wire. Wrap one end of the first four wires to its own nail.

Squeeze the lemons, crushing them gently, to loosen the pulp inside so the juice flows inside.

Insert the nail end of the first wire into the first lemon.

Insert the bare end of the second wire into the first lemon (without letting the wire and the nail touch each other inside the lemon). Insert the nail end of that same wire into the second lemon.

WHAT HAPPENS

The lemons light up the lightbulb.

WHY IT WORKS

The citric acid in the lemon juice acts as an electrolyte, conducting an electron flow between the copper in the wire and the bit of steel in the nail, turning each lemon into a battery.

BIZARRE FACTS

■ The battery owes its discovery to frogs' legs. In the 1780s, Luigi Galvani, a professor of anatomy at Bologna University, noticed that the legs of dead frogs twitched when they were hung from hooks on a rail. Fellow professor Allesandro

Volta of Pavia University deduced that the frogs' legs were completing the circuit between the copper hooks and the iron rail, prompting him to produce a Voltaic cell in 1800.

■ The modern-day household battery should be called a cell, not a battery. A *battery* is an array of single cells.

■ The lemon is actually a type of berry called a *hesperidium*.

■ Lemons are believed to have originated in northeastern India, near the Himalayas.

■ The first lemon trees in America were planted in 1493 by Christopher Columbus.

■ Actor Jack Lemmon's last name really is Lemmon.

■ The word *lemon* is slang for "a defective automobile," derived from the fruit's unavoidable sour taste.

■ The amount of energy needed to make a battery is fifty times greater than the amount of energy that same battery produces.

THE LEMON ELECTRIC CHAIR

Upon learning that the first electric chair had been put to use in Auburn Prison in New York on August 6, 1890, Abyssinian Emperor Memelik II, determined to modernize his country, ordered three electric chairs from the American manufacturer. When the chairs arrived in Abyssinia, Emperor Memelik discovered they would not work without electricity (which his country did not have), so he made one of the electric chairs his royal throne.

WHAT YOU NEED

Wint-O-Green Life Savers

Ziploc storage bag

Wooden block

Hammer

WHAT TO DO

Place one Wint-O-Green Life Saver in the Ziploc bag. Seal the bag and place it on the wooden block. In a dark room or closet, hold the hammer above the Life Saver. Look directly at the Life Saver as you smash it with the hammer.

WHAT HAPPENS

A quick burst of blue-green light flashes the moment the wintergreen candy is crushed.

WHY IT WORKS

Crushing a crystalline substance, in this case the synthetic wintergreen—methyl salicylate—emits light. This phenomenon is called *triboluminescence.*

BIZARRE FACTS:

■ Cat urine glows under a black light.

■ In the early 1900s, Cleveland-based chocolate maker Clarence A. Crane developed the idea for white-circle mints, had a pharmaceutical manufacturer produce them on his pill machine, and named them Life Savers because they resembled the flotation device.

- Advertisements for Life Savers have included the corny puns *holesome*, *enjoy-mint*, *refresh-mint*, and *content-mint*.

- Ninety percent of all mints sold today are Life Savers.

- Six billion Life Savers are produced every year.

- In 1970, the Dow Chemical Company introduced the Ziploc storage bag with its patented tongue-in-groove "Gripper Zipper," providing a virtually airtight, watertight seal that revolutionized plastic bags. "Ziploc" is a clever hybrid of the words *zipper* and *lock*—a mnemonic device to remind consumers that the bags zip shut and lock tight.

LIFE-THREATENING LIFE SAVERS?

In a letter published in the *New England Journal of Medicine*, two Illinois physicians, Dr. Howard Edward, Jr., and Dr. Donald Edward, warned that biting a Wint-O-Green Life Saver while in an oxygen tent, operating room, or space capsule could be life-threatening. The *Journal* declared Wint-O-Green Life Savers safe for oxygen tents and gas stations.

fake blood

WHAT YOU NEED

1½ cups Karo Light Corn Syrup

½ cup Rose's Grenadine

Bottle of red food coloring

Three drops blue food coloring

Clean, empty glass jar

Spoon

WHAT TO DO

Mix the corn syrup, grenadine, and food coloring in the jar to produce a nice, deep blood red. Stir well.

WHAT HAPPENS

You've made a fairly realistic prop blood for stage or screen.

WHY IT WORKS

The grenadine improves the flow of the corn syrup "blood" and lets it soak more realistically into clothing and other fabric.

BIZARRE FACTS

- Hershey's Chocolate Syrup was commonly used as fake blood in black-and-white Hollywood movies.

- Lobsters have blue blood.

- The ancient Aztecs used cochineal, a red dye prepared from the dried bodies of female *Cadtylopius coccus*, an insect that lives on cac-

tus plants in Central and South America. Cochineal is still used today in food coloring, medicinal products, cosmetics, inks, and artists' pigments.

- The heart of a shrimp is located in its head.

- Bloody horror movies worth missing: *Blood Orgy of the She Devils* (1972), *Blood Suckers from Outer Space* (1984), *The Blood Spattered Bride* (1974), and *Bloodsucking Pharaohs from Pittsburgh* (1991).

- To simulate bullet shots in movies, the special effects department attaches "squibs"—small, non-metallic explosive charges—beneath the actor's clothing. Latex "blood bags," filled with a bright red, gelatin-based fluid, can be attached to the squibs, which, when detonated, burst the bags and send the fake blood flowing.

■ People who live in high altitudes have up to two quarts more blood than people who live at sea level.

■ English author Bram Stoker based his novel *Dracula* on Vlad Tepes, the cruel prince of Walachia (now a part of Romania). Prince Vlad was nicknamed Dracula, which means "son of the devil" or "son of a dragon" in Romanian.

■ The common vampire bat, which does attack people and drink their blood, is only three inches long.

FILL 'ER UP

In 1970, Warren C. Jyrich, a fifty-year-old hemophiliac undergoing open heart surgery at the Michael Reese Hospital in Chicago,

required nineteen hundred pints of blood. That's roughly the blood of 190 people.

OUR FRIEND, THE BLOODY MARY

The Bloody Mary, invented in 1920 by Ferdinand Petiot, a bartender at Harry's New York Bar in Paris, was originally named "Bucket of Blood" after a club in Chicago. Contrary to popular belief, the Bloody Mary was not named after Mary, Queen of Scots, although she was beheaded by her cousin, Queen Elizabeth I. The red concoction was named after Queen Mary I of England, who was known as "Bloody Mary" for the brutal persecutions she caused the Protestants in an attempt to bring them back to the Roman Catholic faith. During her five-year reign, more than three hundred people were burned at the stake.

fALLing egg

WHAT YOU NEED

Clean, empty glass jar

Water

Food coloring

8-inch-square piece of cardboard

Empty aspirin bottle

Fresh egg

WHAT TO DO

Fill the jar halfway with water and add three drops of food coloring. Place the piece of cardboard over the mouth of the jar. Place the aspirin bottle in the center of the cardboard directly over the center of the jar. Place the egg point first into the aspirin bottle. Hold the jar firmly with one hand, and quickly pull the piece of cardboard straight out from under the aspirin bottle.

WHAT HAPPENS

The aspirin bottle tumbles off and the egg drops into the jar of water.

WHY IT WORKS

As Newton's first law states, "Objects at rest want to remain at rest, and objects in motion want to remain in motion." The stationary egg wants to remain at rest, but once its support is pulled out from under it, gravity pulls the egg straight down into the jar of water.

BIZARRE FACTS

■ Newton's first law also explains how a skilled magician can pull a tablecloth out from under a fully set dinner table.

■ Eggs do not crush under the weight of a mother bird as she sits on the nest because when a force is applied to an egg, the curve of the egg distributes the force over a wide area away from the point of contact.

■ You can spin a hard-boiled egg in place on a flat surface. A raw egg will wobble.

■ In 1979, David Donoghue dropped fresh eggs from a helicopter 650 feet above a golf course in Tokyo, Japan. The eggs remained intact.

- Dale Lyons of Meriden, Great Britain, holds the world record for running while carrying an egg on a spoon. On April 23, 1990, he ran 26 miles 385 yards in 3 hours 47 minutes.

- On March 23, 1991, Meals on Wheels PLUS of Manatee, Florida, held the biggest Easter egg hunt on record in the United States using 120,000 plastic and candy eggs.

THE SQUEEZE ON EGGS

If you hold an egg in your palm and then try to squeeze your hand into a fist, you will not crush the egg.

floating paper clip

WHAT YOU NEED

Scissors

String

Paper clip

Scotch tape

Clean, empty glass jar with a metal lid

Magnet

WHAT TO DO

Cut a piece of string a few inches high and tie one end to the paper clip. Tape the other end to the bottom of the jar. Place the magnet inside the lid. Show the jar to your audience with the clip lying at the bottom of the jar. Screw the lid on the jar, and then turn the jar upside down so the clip hangs from the string. Turn the jar right side up again.

WHAT HAPPENS

The paper clip remains suspended in air.

WHY IT WORKS

The magnet attracts the paper clip, but the string prevents the paper clip from being pulled to the magnet.

BIZARRE FACTS

- Scientists believe that the earth acts like a huge magnet with magnetic north and south poles.

- In 1900, Cornelius J. Brosnan

received a patent for the Konaclip better known today as the paper clip.

- The Indian rope trick (making an ordinary rope rise into the night sky and then climbing up the rope) is done by stretching thin cable some fifty feet off the ground across a valley and slinging another fine cable over it—one end held by an assistant far off in the distance, the other end attached to a small hook that is attached to the rope.

- In the 1890s, French magician Alexander Herrmann created the illusion of levitation by having a "hypnotized" woman lie on a board placed between the back of two chairs, taking away the two chairs, and then passing a hoop around the floating woman. The illusion is created by having an assistant behind a curtain operate a strong frame set with a "gooseneck" (a metal slat curved like the neck of a goose).

■ In an eighteenth-century treatise, Pope Benedict XIV reported that several eyewitnesses, including Pope Urban VIII, had seen St. Joseph of Cupertino rise into the air "when in a condition of ecstatic rapture."

■ In *Belgravia* magazine, Reverend C. M. Davies, a critic of psychics, detailed his own eyewitness account of seeing British psychic Daniel Dunglas Home float around a drawing room for five minutes.

SEPARATE TALES OF ABSURDITY

In his nonfiction books, anthropologist Carlos Castaneda reported seeing Mexican Indian sorcerer Don Genaro fly through the air, walk horizontally up the side of a tree, and fly back. He also claimed to have seen both Don Genaro and Don Juan jump off a cliff, twirl slowly in the air, reach bottom, and float back up to the top. Many consider Castaneda's nonfiction books to be works of novelistic fancy.

floating ping-pong ball

WHAT YOU NEED
Small funnel
Ping-Pong ball

WHAT TO DO
Turn the funnel upside down. Hold the Ping-Pong ball in the funnel with one finger. Fiercely blow down into the narrow end of the funnel, then let go of the ball and continue blowing.

WHAT HAPPENS
The ball floats inside the funnel.

WHY IT WORKS
According to Bernoulli's principle (named for Swiss mathematician Daniel Bernoulli), the air blowing rapidly over the top of the ball creates increased air pressure under the ball, which then holds the ball up by air.

BIZARRE FACTS
- Bernoulli's principle explains how airplanes fly. The top of an airplane wing curves so that air flowing over it speeds up, moving faster than the air flowing below the wing.
- Bernoulli's principle also explains how baseball pitchers throw

curve balls. The pitcher spins the ball as it is thrown. The spin increases the speed of the air flowing over one side of the ball, and the disproportionate air pressure causes the ball to curve from its original course.

- Golf balls always spin backward when struck, which, if the balls were smooth, would create higher air pressure above the ball, preventing it from traveling more than seventy yards. Dimples in the ball carry air upward over the top, creating low air pressure over the ball, allowing it to be driven up to three hundred yards.

- While the name *Ping-Pong* sounds Chinese, the game originated in England during the late 1800s.

- The word *Ping-Pong*, a trademark for table tennis equipment, is derived from the sound the celluloid ball makes when it hits the paddle and then the table.

- In the movie *Forrest Gump*, Forrest, played by Tom Hanks, becomes a self-taught Ping-Pong champion and competes in the first Sino-American table tennis tournament.

- Ping-Pong became an Olympic sport in 1988. Shuffleboard has yet to gain Olympic status.

- One of the many elaborate gifts Richard Burton purchased for Elizabeth Taylor was the $38,000 "Ping-Pong" diamond.

NOT-SO-LUCKY LINDY

Charles Lindbergh was not the first person to fly nonstop across the Atlantic Ocean. That distinction is actually held by two British aviators, John Alcock and Arthur Whitten Brown, who flew from Newfoundland to Ireland in 1919. Eight years later, Lindbergh became the first person to fly nonstop across the Atlantic Ocean *alone*.

flying rice krispies

WHAT YOU NEED

Two strips of stiff 1-by-3-inch cardboard

Pencil

Scissors

Glass chimney from a hurricane lamp

Strip of 4-by-11-inch poster board

Scotch tape

½ cup Rice Krispies cereal

Blow dryer

WHAT TO DO

Holding the 1-by-3-inch cardboard strips horizontally, draw a line through the center of each strip, perpendicular to its length. Cut ½ inch into each line. Fit the two cardboard strips together at the cuts to form a sturdy, cross-shaped support for the glass chimney.

Set the glass chimney on top of the cardboard support. Tape the ends of the 4-by-11 inch strip of poster board together to form a loop. Place the loop around the glass chimney's cardboard support to create a fence.

Pour the Rice Krispies into the opening at the top of the chimney.

Use the blow dryer to blow a stream of air across the top of the glass chimney.

WHAT HAPPENS

The Rice Krispies rise and float through the glass chimney.

WHY IT WORKS

The air blowing rapidly across the top of the glass chimney creates increased air pressure at the bottom of the glass chimney, causing the Rice Krispies to rise. Bernoulli's principle puts it this way: As the velocity of a gas or liquid increases, the pressure perpendicular to its direction of flow decreases.

BIZARRE FACTS

■ In the 1890s, the health-conscious Dr. John Harvey Kellogg and his brother, W. K. Kellogg, invented Corn Flakes while working at the Battle Creek Sanitarium, and had a spiteful relationship for the rest of their lives.

- The Kellogg brothers invented peanut butter, but failed to patent it.

- C. W. Post, a former patient of the Kellogg brothers at the Battle Creek Sanitarium, launched Grape Nuts, a cereal similar to granola, a cereal invented by the Kelloggs and served at the sanitarium.

- Although he invented granola and Corn Flakes, Dr. John Harvey Kellogg breakfasted daily on seven graham crackers.

- The average American cupboard contains four different brands of cereal.

- In 1959, with only $100, Leandro P. Rizzuto and his parents, Julian and Josephine, started Continental Hair Products in New York City to market hair rollers for beauty salons. In 1968, the company developed the hot comb, and in 1971 introduced the first hand-held pistol-grip blow dryer.

- Bernoulli's principle makes it possible to sail a boat forward against the wind.

SNAP! CRACKLE! POP!

In 1928, the Kellogg Company introduced Kellogg's Rice Krispies as "the Talking Cereal." In 1933, a year after the phrase "Snap! Crackle! Pop!" was printed on the box, a tiny, nameless gnome wearing a baker's hat appeared on the side of the box. He eventually became known as Snap, and in the mid-1930s he was joined by Crackle (wearing a red striped stocking cap) and Pop (in a military hat). Snap, Crackle, and Pop are the Kellogg Company's oldest cartoon characters.

fried marbles

WHAT YOU NEED
Bag of transparent marbles

Frying pan

Wooden spoon

Oven mitt

Large pot

Cold water

Ice

WHAT TO DO
With adult supervision, place the marbles in the frying pan, and, wearing the oven mitt, heat them over a high heat, stirring with the wooden spoon.

Fill the large pot with cold water, add the ice, and let cool.

When the marbles are piping hot, pour them into the ice water. Let them cool off, then dry.

WHAT HAPPENS
The glass inside the marbles shatters into shards and looks like shimmering crystal.

WHY IT WORKS
When glass goes from extreme heat to extreme cold, it cracks from the inside out.

BIZARRE FACTS
■ Although marbles have been made from clay, stone, wood, glass, and

steel, most marbles today are made from glass.

- Marbles found in ancient Egypt and Rome can be seen in the British Museum.

- When a German glassblower invented a tool called the marble scissors in 1846, the manufacture of glass marbles became economically feasible.

- World War I cut off the supply of marbles to North America.

- Most glass marbles in the United States are made at a plant in Clarksburg, West Virginia, which makes millions every year.

- A variety of colors and intricate patterns create a wide range of glass marbles, including the Immy, Moonstone, Rainbow, Marine, Cat's Eye, Genuine Carnelian, First American, Japanese Cat's Eye, Scrap Glass, and Peppermint Stripe.

- Marble games include Archboard, Bounce About, Bounce Eye, Conqueror, Die Shot, Dobblers, Eggs in the Bush, Handers, Hundreds, Increase Pound, Lag Out, Long Taw, Odds or Even, One Step, Picking Plums, Pyramid, Ring Taw, Spanners, and Three Holes.

- Marbles are often used at the bottom of clear glass vases to support flowers or at the bottom of fish tanks.

LOSING YOUR MARBLES?

The most common method of shooting a marble is known as *fulking*.

green ſlime

WHAT YOU NEED

4-ounce bottle of Elmer's Glue-All

Two large glass bowls

Water

Green food coloring

Large spoon

Measuring cup

1 teaspoon 20 Mule Team Borax

Ziploc storage bag or airtight container

WHAT TO DO

Empty the bottle of Elmer's Glue-All into the first bowl. Fill the empty glue bottle with water and then pour it into the bowl of glue. Add ten drops of food coloring and stir well.

In the second bowl, mix the borax with 1 cup water. Stir until the powder dissolves.

Slowly pour the colored glue into the bowl containing the borax solution, stirring as you do so. Remove the thick glob that forms, and knead the glob with your hands until it feels smooth and dry. Discard the excess water remaining in the bowl. Store the green slime in the Ziploc bag or airtight container.

WHAT HAPPENS

The resulting soft, pliable, rubbery glob snaps if pulled quickly, stretches if pulled slowly, and slowly oozes to the floor if placed over the edge of a table.

WHY IT WORKS

The polyvinylacetate molecules in the glue act like invisible bicycle chains drifting around the water. The borax molecules (sodium tetraborate) act like little padlocks, locking the chain links together wherever they touch the chain. The locks and chains form an interconnected "fishnet," and the water molecules act like fish trapped in the net.

BIZARRE FACTS

■ Green slime is a non-Newtonian fluid—a liquid that does not abide by any of Sir Isaac Newton's laws on how liquids behave. Quicksand, gelatin, and ketchup are all non-Newtonian fluids.

■ Increasing the amount of borax in the second bowl makes the slime thicker. Decreasing the amount of borax makes the slime more slimy and oozy.

■ A non-Newtonian fluid's ability to flow can be changed by applying a force. Pushing or pulling on the slime makes it temporarily thicker and less oozy.

■ 20 Mule Team Borax is named for the twenty-mule teams used during the late nineteenth century to transport borax 165 miles across the desert from Death Valley to the nearest train depot in Mojave, California. The twenty-day round trip started 190 feet below sea level and climbed to an elevation of over 4,000 feet before it was over.

■ Between 1883 and 1889, the twenty-mule teams hauled more than twenty million pounds of borax out of Death Valley. During this time, not a single animal was lost nor did a single wagon break down.

■ Today it would take more than 250 mule teams to transport the borax ore processed in just one day at Borax's modern facility in the Mojave desert.

■ Although the mule teams were replaced by railroad cars in 1889, twenty-mule teams continued to make promotional and ceremonial appearances at events ranging from the 1904 St. Louis World's Fair to President Woodrow Wilson's inauguration in 1917. They won first place in the 1917 Pasadena Rose Parade and attended the dedication of the San Francisco Bay Bridge in 1937.

■ Borax deposits in Death Valley were abandoned when richer deposits were found elsewhere in the Mojave desert, turning mining settlements into ghost towns that now help make the region a tourist attraction.

- According to legend, borax was used by Egyptians in mummification.

- In the furniture business, the word *borax* signifies cheap, mass-produced furniture.

- 20 Mule Team Borax was once proclaimed to be a "miracle mineral" and was used to aid digestion, keep milk sweet, improve the complexion, remove dandruff, and even cure epilepsy.

WHO YA GONNA CALL?

The 1984 movie *Ghostbusters*, starring Bill Murray, Dan Aykroyd, Sigourney Weaver, and Harold Ramis, and featuring ghosts that spewed slime, inspired the catchphrase "I've been slimed."

THE ELMER STORY

In 1936, Borden launched a series of advertisements featuring cartoon cows, including Elsie, the spokescow for Borden dairy products. In 1940, compelled by Elsie's popularity, Borden dressed up "You'll Do Lobelia," a seven-year-old, 950-pound Jersey cow from Brookfield, Massachusetts, as Elsie for an exhibit at the World's Fair. She stood in a barn boudoir decorated with whimsical props including churns used as tables, lamps made from milk bottles, a wheelbarrow for a chaise lounge, and oil paintings of Elsie's ancestors—among them Great Aunt Bess in her bridal gown and Uncle Bosworth, the noted Spanish-American War admiral. This attracted the attention of RKO Pictures, which hired Elsie to star with Jack Oakie and Kay Francis in the movie *Little Men*. Borden needed to find a replacement for Elsie for the World's Fair exhibit. Elsie's husband, Elmer, was chosen, and the boudoir was converted overnight into a bachelor apartment, complete with every conceivable prop to suggest a series of nightly poker parties. In 1951, Borden chose Elmer to be the marketing symbol for all of Borden's glue and adhesive products. Elsie the Cow and her husband, Elmer, have two calves, Beulah and Beauregard.

Homemade Ufo

WHAT YOU NEED

Eight sheets of thin tissue paper

Paper clips

Pencil

Scissors

Elmer's Glue-All

Strip of poster board ¼ inch wide by 28 inches long

Thin iron wire

Tin snips

Clean, empty coffee can

Metal file

Wad of absorbent cotton, about the size of a baseball

Rubbing alcohol

Long kindling match

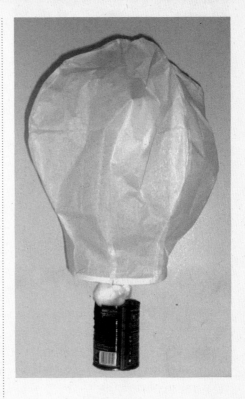

WHAT TO DO

Fold the sheets of tissue paper in half lengthwise. Secure the sheets together along the fold with several paper clips. With the pencil, reproduce the pictured pattern on the top sheet of tissue paper, and use the scissors to cut it out through all sixteen sheets.

With the first panel lying vertically on the table, apply glue only along the right edge from top to bottom. Lay the second panel on top of the first and press the glued edge. Apply glue along the left edge of the second panel, lay the third panel on top of it, and press the glued edges together. Finish gluing the remaining panels in this accordion style, then unfold the balloon and glue the edge of the last panel to the edge of the first panel, sealing the balloon. Trim the excess tissue at the top and glue a small piece of tissue on to seal the top.

Glue the ends of the poster board strip together to form a loop. Let dry.

Apply glue to the loop's outside edge, insert the loop inside the bottom lip of the tissue paper balloon, and seal in place, crimping or trimming any excess tissue, if necessary.

Cut two pieces of thin iron wire eleven inches in length. Tie the two

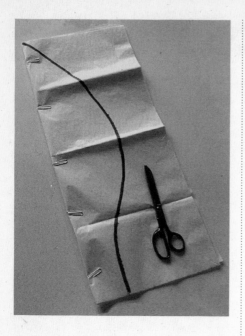

pieces of wire from side to side so the wires intersect to form a cross in the center of the loop. (You can use an unfolded paper clip or a sewing needle to puncture holes in the loop to tie the wires in place.)

With adult supervision use the tin snips to carefully cut a 3-by-5-inch window in the coffee can, starting at the lip of the can. Use the metal file to file the sharp edges.

Place the coffee can on the ground, open the balloon, and have an assistant hold the balloon on the can so it rests level.

Wrap a small amount of wire around the cotton wad, leaving enough wire to protrude to form a small hook. Moisten (not saturate)

the cotton wad with rubbing alcohol, then seal the bottle of alcohol and move it to a safe place far from the balloon. Insert the cotton wad through the window of the coffee can and attach its wire hook to the criss-crossed wires in the center of the frame, so it hangs from the balloon into the coffee can.

With adult supervision and using the kindling match, carefully light the cotton, without igniting the tissue paper or posterboard loop.

WHAT HAPPENS

As hot air rises inside the tissue-paper balloon, the balloon fully inflates and rises from its coffee can launch pad. The balloon drifts upward until the flame goes out and air inside the balloon cools. If launched at night, the flame also illuminates the balloon, making it look like a UFO.

WHY IT WORKS

Heat rises. The hot air inside the balloon, lighter than the surrounding cool air, causes the balloon to rise.

BIZARRE FACTS

■ Weather balloons are commonly mistaken for UFOs.

■ The Bible includes the prophet Ezekiel's eyewitness account of an aerial chariot containing four-winged monsters and traveling on wheels within wheels in a stormy wind with flashes of fire. UFO enthusiasts believe Ezekiel was describing an alien spacecraft. Biblical scholars point out that Ezekiel describes each monster as having four faces unique to this planet: the face of a man, a lion, an ox, and an eagle.

■ When viewed from the sky, the Nazca lines—the gigantic markings made centuries ago on the plains of Nazca, Peru—can be clearly seen as enormous drawings of animals. While UFO enthusiasts suggest that these fantastic drawings were meant as signals for extraterrestrial visitors, archaeologists have shown that these drawings were ancient astronomers' method of tracking the constellations for agricultural purposes.

■ Seven percent of all Americans claim to have seen a UFO.

■ In the liner notes to his album *Walls and Bridges*, John Lennon claimed: "On the 23rd Aug. 1974 at 9 o'clock, I saw a U.F.O."

■ If NASA sent birds into space, they would die. Birds need gravity to swallow.

- The "live long and prosper" hand gesture made by Mr. Spock on *Star Trek* is the hand gesture used by Jewish priests (*kohanim*) while saying certain prayers.

- David Prowse, the actor dressed as Darth Vader in *Star Wars*, spoke all of Vader's lines as the movie was filmed. He did not know until he saw a screening of the movie that his voice had been dubbed over by James Earl Jones.

- In the movie *The Wizard of Oz*, the Wizard claims "I'm an old Kansas man myself," yet he takes off from the Emerald City in a hot-air balloon painted with the words "Omaha State Fair," which could only take place in Nebraska— proving that he too is filled with hot air.

UNIDENTIFIED FLYING BUTTON

UFO hoaxes are often perpetrated by trick photography. A photo of an alleged UFO taken by an airline pilot over Venezuela in 1965 fooled experts for six years—until an engineer in Caracas admitted that he had taken a photograph of a button, placed it over an aerial shot, rephotographed it, and burned in a shadow of the UFO during processing.

Hoſe pHONe

WHAT YOU NEED

Hacksaw

100-foot rubber garden hose

Two large funnels

Black electrical tape

WHAT TO DO

Use the hacksaw to cut off the metal couplers at the ends of the garden hose. Insert the narrow end of a funnel into each end of the hose and secure in place with the black electrical tape. Hold one funnel to your ear and the second funnel to your mouth. Say something. You will hear your voice with a slight delay. Stretch the hose across the yard and have a friend speak into one funnel while you listen with the second funnel.

WHAT HAPPENS

The hose works just like a phone, with a slight delay.

WHY IT WORKS

At sea level, sound waves travels 760 miles per hour. That means sound travels through a 100-foot-long hose in one tenth of a second.

BIZARRE FACTS

■ Bats make high-frequency sounds while flying and navigate using the echoes from these sounds to determine the distance and direction of objects in the area. Using reflected sound waves to navigate is called *echolocation*.

■ In 1860, the inventor of vulcanized rubber, Charles Goodyear, died, having failed to perfect a practical use for his invention and leaving his family with nearly $200,000 in debts. Ten years later, Dr. Benjamin Franklin Goodrich, determined to cash in on rubber's untapped potential, founded the B. F. Goodrich Company in Akron, Ohio, and began producing the world's first rubber hoses.

■ The 1970s television sitcom *Welcome Back, Kotter*, starring Gabe Kaplan and John Travolta, popularized the meaningless catchphrase "Up your nose with a rubber hose."

- Around 350 B.C.E., the Greek philosopher Aristotle suggested that sound is carried to our ears by the movement of air. He was wrong. Sound travels in waves.

- The higher the altitude, the slower sound travels.

- Two days before he broke the sound barrier, Chuck Yeager broke two ribs.

- On Alexander Graham Bell's telephone, the mouthpiece also served as the earpiece. Thomas Edison separated the transmitter from the receiver, making the telephone easier to use.

TELEGRAM ABOUT THE TELEPHONE

In 1876, an internal memo at Western Union read: "This 'telephone' has too many shortcomings to be seriously considered as a means of communication. The device is inherently of no value to us."

HUMAN Lightbulb

WHAT YOU NEED
Wool sweater

Fluorescent lightbulb

WHAT TO DO
Put on the wool sweater. In a dark room, rub the fluorescent lightbulb briskly against the sweater.

WHAT HAPPENS
The fluorescent lightbulb glows.

WHY IT WORKS
The friction creates a static charge strong enough to cause the gas inside the tube to fluoresce.

BIZARRE FACTS
- On the 1960s television show *The Addams Family*, electrically charged Uncle Fester makes a lightbulb illuminate by simply placing it in his mouth.

- The fluorescent bulb is more economical and energy efficient than the incandescent bulb, which wastes up to 80 percent of its energy generating heat.

- While Thomas Edison is credited with inventing the first incandescent lamp using carbon for the filament, English inventor Joseph Swan patented his incandescent lamp using carbon for the filament in 1878, a year before Edison.

- When cartoon characters get an idea, an incandescent lightbulb goes off over their heads—never a fluorescent bulb.

- Thomas Edison, nicknamed "the Wizard of Menlo Park," was expelled from school in Port Huron, Michigan, after the schoolmaster incorrectly diagnosed him as being mentally retarded. Edison was actually partially deaf, the result of a bout with scarlet fever.

- Lightning travels between a hundred and a thousand miles per second, generating a temperature up to 54,000°F, six times hotter than the surface of the sun.

ENLIGHTENMENT

American park ranger Roy Sullivan was struck by lightning seven times between 1942 and 1977.

invisible inks

LEMON JUICE OR MILK

WHAT YOU NEED

Q-tips cotton swab

Lemon juice or milk

Paper

Electric iron

Ironing board

WHAT TO DO

Using the Q-tips cotton swab as a paintbrush and the lemon juice or milk as ink, write your message on the paper. Let dry at room temperature. With adult supervision, iron the piece of paper.

WHAT HAPPENS

The invisible message appears.

WHY IT WORKS

Lemon juice and milk are simple organic liquids that appear invisible once they have dried on a sheet of paper, but darken when held over a heat source.

CORNSTARCH

WHAT YOU NEED

1 teaspoon cornstarch

Measuring cup

Water

Microwave-safe container

Spoon

Q-tips cotton swab

Paper

Two small bowls

Iodine

Two small sponges

Eyedropper

Lemon juice

WHAT TO DO

Mix the cornstarch and ½ cup water in the microwave-safe container. Stir until smooth. Heat in the microwave on high for fifteen seconds, stir, then heat on high for forty-five seconds more.

Dip the Q-tips cotton swab into the mixture. Write your message on the paper. Let dry.

In one small bowl, add ten drops of iodine to ¼ cup water. Sponge the iodine solution lightly over the paper.

WHAT HAPPENS

The message appears in dark blue. (If the paper contains starch, the paper may also turn light blue.)

In the second bowl, pour ¼ cup lemon juice. Sponge it over the message. The message will disappear again.

WHY IT WORKS

The iodine reacts with the corn-starch to form a new compound that appears blue-black. The ascorbic acid in lemon juice combines with the iodine to form a colorless compound.

BIZARRE FACTS

■ Prisoners of war used their own sweat and saliva as invisible ink.

■ India ink was actually invented in China.

■ During World War II, a German spy named Oswald Job was executed in 1944 after a crystal of concentrated invisible ink was found hidden in a hollow key.

■ A message in invisible ink, written beneath a one-sentence love letter on a postcard sent from Poland in 1943, vividly describes the horrific conditions in a Nazi death camp and makes an urgent request for supplies. The postcard is on display at the Yad Vashem Holocaust Museum in Jerusalem.

■ In Mexico, Parker Pen's advertising agency translated the phrase "won't leak in your pocket and embarrass you" by incorrectly using the Spanish word *embarazar*, which made the phrase read, "won't leak in your pocket and make you pregnant."

CLOSE SHAVE

In 500 B.C.E., Histiaeus wrote a secret message on the shaved head of a slave, waited for his hair to grow back, and then sent the slave across enemy lines. When the slave's head was shaved, the message was revealed.

killer ʃtraw

WHAT YOU NEED

Raw potato

Two plastic drinking straws

WHAT TO DO

Place the potato on a table. Hold the first straw at the top (without covering the hole) and try to stab it into the potato. Hold your thumb over the hole in the top of the second straw and try to stab it into the potato.

WHAT HAPPENS

The open-ended straw bends, and only a little bit of the straw penetrates the potato. The closed straw cuts deeply into the potato.

WHY IT WORKS

The air trapped inside the straw gives the straw enough strength to penetrate the skin of the potato. As the straw enters the potato, the potato plug compresses the air inside the straw, increasing the air pressure, and strengthening the straw.

BIZARRE FACTS

■ During colonial times, New Englanders, convinced that raw potatoes contained an aphrodisiac that induced behavior that shortened a person's life, fed potatoes to pigs as fodder.

■ After serving as ambassador to France, Thomas Jefferson brought the recipe for French-fried potatoes to America, where he served them to guests at Monticello, popularizing French fries in the United States.

■ The potato originated in South America, where the Incas cultivated and crossbred it. In the 1500s, while Spanish explorers introduced the potato to Europe, English explorers brought the potato to the United Kingdom, where it became the principal crop of Ireland. Today, Russia grows nearly 30 percent of the world's potatoes, more than any other country.

■ Dan Quayle, vice-president of the United States under George Bush, corrected a student in a spelling bee, insisting that the word *potato* is spelled *potatoe*. It isn't.

■ George Gershwin wrote the famous lyric, "You like potato, and I like po-tah-to" in his 1937 hit song "Let's Call the Whole Thing Off."

■ The phrases "that's the last straw" and "the straw that the broke the camel's back," originated with Archbishop John Bramhall, who, in 1655, wrote: "It is the last feather that breaks the horse's back."

■ Straw and hay are not the same thing. Straw is the dried stems of grains such as wheat, rye, oats, and barley. Hay is dried grasses or other plants.

SMALL POTATOES

When Pope John Paul II visited Miami, Florida, a local T-shirt maker translated the phrase "I saw the Pope" into Spanish by incorrectly using the feminine *la papa* instead of the masculine *el papa*, which made the T-shirts read: "I saw the potato."

GENUINE LAVA LAMP

WHAT YOU NEED

Rubber gloves

Colored Sharpie permanent marker

Bottle of mineral oil (from a drug-store)

Food coloring

Bottle of 91 percent isopropyl alcohol (from a drugstore)

Bottle of 70 percent isopropyl alcohol (from a drugstore)

Funnel

Clean, empty 1-quart glass bottle with cap

Can opener

Empty coffee can

Tin snips

40-watt lightbulb

Ceramic lightbulb base

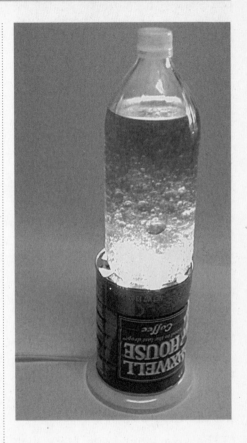

WHAT TO DO

Wearing rubber gloves, remove the ink-soaked felt tube from the inside of the Sharpie permanent marker and place it inside the full bottle of mineral oil. Let it sit for an hour until the mineral oil takes on the color of the Sharpie ink.

Put five drops of food coloring (a different color from the Sharpie ink) in the bottle of 91 percent isopropyl alcohol and five drops of the same food coloring in the bottle of 70 percent isopropyl alcohol.

Using the funnel, pour the colored 90 percent alcohol into the glass bottle. Pour the colored mineral oil into the bottle. The mineral oil will sink to the bottom. Slowly add the colored 70 percent isopropyl alcohol into the bottle (gently mixing as you do) until the mineral oil ("lava") seems lighter and is about to jump off the bottom. Use the two alcohols to adjust the responsiveness of the lava. Seal the bottle tightly.

With the can opener, punch eight holes around the bottom of the coffee can. Using the tin snips, cut a hole 2 inches in diameter in the center of the bottom of the coffee can. Place the lightbulb in the ceramic lightbulb base and cover with the upside-down coffee can.

Place the sealed bottle on top of the coffee can and over the lightbulb. Plug in the lightbulb and observe.

WHAT HAPPENS

The lava forms large blobs that repeatedly rise to the surface and fall back to the bottom of the glass bottle.

WHY IT WORKS

Oil and alcohol are *immiscible*—meaning they do not mix, but instead separate into layers. The heat of the lightbulb warms the oil, making it lighter than the alcohol, so it rises. When the oil reaches the top of the mixture, it cools and settles back down to the bottom of the bottle, only to be reheated.

POOR MAN'S LAVA LAMP

WHAT YOU NEED
Bottle of club soda
Four or five raisins

WHAT TO DO

Open the bottle of club soda, drop in the raisins, and reseal the cap tightly on the bottle. (The effect will also work in a drinking glass filled with club soda.)

WHAT HAPPENS

The raisins sink to the bottom of the bottle, slowly rise to the surface, and sink back down again, repeatedly.

WHY IT WORKS

Carbon dioxide bubbles from the club soda accumulate in the wrinkles of the raisins, eventually lifting the raisins to the surface. There the bubbles escape, and the raisins sink to the bottom again to repeat the cycle. The effect lasts longer in a sealed bottle of club soda because less carbon dioxide is able to escape.

BIZARRE FACTS

■ Craven Walker, a native of Singapore, came up with the idea for the lava lamp while drinking in an English pub after World War II, and spent the next fifteen years developing it, just in time for the psychedelic sixties.

■ Over four hundred thousand lava lamps are made each year.

■ Firewalkers in Hawaii walk barefoot over hot lava.

■ The eruption of Laki volcano in 1783 in southeast Iceland created a lava flow some 43½ miles long, the longest in recorded history.

■ Vinegar and oil, the main ingredients in mayonnaise, will not dissolve in one another. But a third ingredient—egg yolks—contain several *emulsifiers*, which form a protective skin around the beaten oil droplets, preventing them from mixing—merging just like the wax in a lava lamp.

■ Authentic lava lamps contain a specially compounded wax and

liquid composed of eleven secret ingredients.

■ Baby oil is made primarily from mineral oil, a clear, colorless, oily liquid with little odor or taste. Distilled from petroleum, mineral oil is used medicinally as a lubricant and laxative, as an oil base in cosmetics and hair tonics, and as an ingredient of paints and varnishes.

HOT STUFF

Lava, the molten rock that pours out of volcanoes, reaches temperatures up to ten times hotter than boiling water.

Magic Crystal Garden

WHAT YOU NEED

Five charcoal briquets

Hammer

2-quart glass bowl

Clean, empty glass jar

1 tablespoon ammonia

6 tablespoons salt

6 tablespoons Mrs. Stewart's Liquid Bluing (from the grocery store, or call 1-800-325-7785, or visit http://www.mrsstewart.com)

2 tablespoons water

Food coloring

WHAT TO DO

Break up the charcoal briquets into 1-inch chunks with the hammer and place the pieces in the bottom of the bowl.

In the jar, mix the ammonia, salt, bluing, and water. Mix well.

Pour the mixture over the charcoal in the bowl.

Sprinkle a few drops of food coloring over each piece of charcoal.

Let the bowl sit undisturbed in a safe place for seventy-two hours.

WHAT HAPPENS

Fluffy, fragile crystals form on top of the charcoal, and some climb

up the sides of the bowl. To keep the crystals growing, add another batch of ammonia, salt, bluing, and water.

WHY IT WORKS

As the ammonia speeds up the evaporation of the water, the blue ion particles in the bluing and the salt get carried up into the porous charcoal, where the salt crystallizes around the blue particles as nuclei. These crystals are porous, like a sponge, and the liquid below continues to move into the openings and evaporate, leaving layers of crystals.

BIZARRE FACTS

- All solids have an orderly pattern of atoms, which is repeated again and again. This orderly pattern, called *crystallinity*, can be seen in simple crystals because their shapes reveal their particular atomic structure to the naked eye.

- Some New Age enthusiasts believe that wearing a crystal—usually an amethyst, rose quartz, or clear quartz—around the neck attracts good vibrations and can be used to better arrange a person's spiritual and physical energies. There is no scientific evidence to support this superstition.

- Crystals grow by attracting the atoms of similar compounds, which join together in an orderly pattern. Impure atoms can invade the atomic structure of the crystal and create mixed crystals of dazzling hues.

- In the fifteenth century, amethyst was believed to cure drunkenness.

- Some scientists theorize that birds have a tiny magnetic crystal in their brain, enabling them to navigate during migration by detecting the earth's magnetic field.

- Crystal gardens became popular during the Depression and are still known to some as a "Depression flower" or "coal garden."

- The word *crystal* is derived from the Greek word *kyros*, meaning "icy cold." Rock crystal, a colorless quartz, was believed to be ice that had frozen so cold it would never melt.

- In 1921, Henry Ford, eager to find a use for the growing piles of wood scraps from the production of his Model Ts, learned of a process for turning the wood scraps into charcoal briquets, and one of his relatives, E. G. Kingsford, helped select the site for Ford's charcoal plant. In 1951, Ford Charcoal was renamed Kingsford Charcoal.

■ In May 1959, the United States sent two young female monkeys, Able and Baker, into space in a Jupiter rocket. Monkey Able, dressed in a space suit, wore gauze and charcoal diapers.

■ More than 77 percent of all households in the United States own a barbecue grill. Nearly half of those grill owners barbecue year round and, on the average, use their grills five times a month.

■ Mrs. Stewart's bluing, a very fine blue iron powder suspended in water, optically whitens white fabric. It does not remove stains or clean the fabric, but merely adds a microscopic blue particle to white fabric that makes it appear whiter. The brightest whites have a slight blue hue

that, unfortunately, washes out over time. Adding a little diluted bluing to the rinse cycle gives white fabrics this blue hue again, making them look snow-white.

■ Because blue-white is the most intense white, most artists when portraying a snow scene will use the color blue to intensify the whiteness.

■ Some pet owners use Mrs. Stewart's Bluing to whiten white pet fur.

■ Freshly cut carnations placed in a vase with a high content of Mrs. Stewart's Bluing in the water will by osmosis carry the blue color into the tips of the petals quickly.

CHARCOAL ON THE MOON

On July 20, 1969, Neil Armstrong, the first man on the Moon, spoke the first words on the Moon: "That's one small step for man, one giant leap for mankind." The second thought he expressed was: "The surface is fine and powdery. It adheres in fine layers, like powdered charcoal, to the soles and sides of my foot."

mAgNetic bAlLooN

WHAT YOU NEED
Handful of Rice Krispies
Balloon
Wool sweater

WHAT TO DO
Pour the Rice Krispies on a table-top. Inflate the balloon and knot it. Rub the balloon against the wool sweater, about five strokes. Hold the balloon an inch above the Rice Krispies.

WHAT HAPPENS
The Rice Krispies hop up and stick to the balloon.

WHY IT WORKS
The balloon rubs electrons off the wool sweater, giving the balloon a negative charge. The negative charges on the balloon attract the positive charges of the Rice Krispies, overcoming the force of gravity.

BIZARRE FACTS
■ The Kellogg Company is the world's largest maker of ready-to-eat cereals, selling nearly one out of every three boxes of cereal in the United States.

■ The Kellogg Company makes twelve of the top fifteen cereals in the world, including All-Bran, Froot Loops, Kellogg's Corn Flakes, Rice Krispies, and Special K.

■ In 1986, the Kellogg Company stopped giving tours of its Battle Creek, Michigan, factory to prevent industrial spies from unearthing its secret recipes.

- In front of the Kellogg factory in Battle Creek stands a giant statue of Tony the Tiger.

- When W. K. Kellogg refused to buy his grandson's process for making puffed corn grits (developed on company time), the younger Kellogg—John L. Kellogg—started his own company to make the new cereal. His grandfather sued, and John Kellogg committed suicide in his Chicago factory.

KELLOGG'S LUCKY NUMBER

W. K. Kellogg was obsessed with the number seven. Born the seventh son to a seventh son on the seventh day of the week on the seventh day of the month with a last name seven letters long, Kellogg always booked hotel rooms on the seventh floor and insisted that his Michigan license plates end in a seven—which fails to explain the introduction of Kellogg's Product 19. It is not known whether Kellogg drank 7-Up.

marbled paper

WHAT YOU NEED

Newspaper

Baking pan

Water

2 tablespoons white vinegar

Mortar and pestle

Six sticks of colored chalk (different colors)

Six paper cups

Six tablespoons cooking oil

Spoon

Half sheets of white paper

Paper towels

Add a tablespoon of oil to each cup, stirring thoroughly with the spoon. Pour the contents of each paper cup into the pan of water. The chalky colored oil will form large pools on the water's surface.

Gently lay a piece of paper on the water's surface for a moment, lift off, then set to dry on a sheet of newspaper for twenty-four hours. When the marbled paper dries, gently wipe off any surface chalk grains with a paper towel.

WHAT HAPPENS

Swirling patterns of colored oil stick to the paper.

WHY IT WORKS

The molecules of chalk (calcium carbonate), vinegar (acetic acid), water, and the surface of the paper all chemically combine, causing a chemical bond that makes the swirling colors stick to the paper.

WHAT TO DO

Cover the kitchen counter with newspaper. Fill the pan with water, add the vinegar, and place the pan in the middle of the newspaper.

Using the mortar and pestle, crush a piece of colored chalk to a fine powder, then pour into a paper cup. Repeat for all six pieces of chalk, using a different cup for each piece.

BIZARRE FACTS

- Marbling paper was practiced in Japan and China as early as the twelfth century. According to a Japanese legend, the gods gave knowledge of the marbling process to a man named Jiyemon Hiroba as a reward for his devotion to the Katsuga Shrine.

- For centuries, paper marbling masters worked in secrecy to maintain a shroud of mystery to prevent others from mastering the craft and going into business for themselves.

- At age twenty-three, the Italian artist Michelangelo carved the *Pietà*, a statue now in St. Peter's Church depicting the Virgin Mary cradling Jesus after his crucifixion, from a block of marble quarried from Carrara, Italy.

paper Helicopter

WHAT YOU NEED

Sheet of 8½-by-11-inch paper

Scissors

Ruler

Pencil

Two paper clips

WHAT TO DO

Fold the paper in half lengthwise and cut along the fold with the scissors.

Fold one of the halves in half lengthwise.

From the base of the paper, measure 4 inches up the length of the folded sheet of paper and draw a line across the width of the paper. From the base of the paper, measure 6 inches up the length of the folded sheet of paper and draw a line across the width of the paper.

On the line drawn at 4 inches, measure 1 inch in from the open edge of the folded paper. Draw a diagonal line from this point to the point where the line drawn at 6 inches touches the open edge of the folded paper.

Cut out the triangle, cutting through both layers of the paper.

Open the paper and cut the center fold from the top of the paper to the line drawn at 6 inches.

Fold the tabs at the bottom of the paper toward the center and attach two paper clips to the bottom.

Fold the wings in opposite directions along the line drawn at 6 inches.

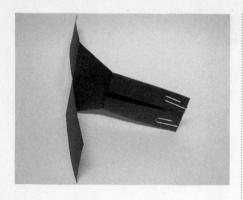

Hold the helicopter above your head and drop it.

WHAT HAPPENS

The paper helicopter rotates like a real helicopter, and the more paper clips you add, the faster the helicopter rotates.

WHY IT WORKS

As the helicopter falls, air rushes out from under the wings in all directions. The air hits the body of the craft, causing it to rotate around a central point. Adding more paper clips increases the weight and reduces the air resistance but increases the amount of air hitting the helicopter wings.

BIZARRE FACTS

■ If you are north of the equator, the paper helicopter spins clockwise when the left wing is bent toward you and counterclockwise when the right wing is bent toward you. If you are south of the equator, the opposite is true.

■ If you flush the toilet north of the equator, the water goes down the drain counterclockwise. South of the equator, the water goes down the drain clockwise.

■ Although Russian-born Igor Sikorsky is credited with inventing the first successful single-rotor helicopter in 1939, French inventor Paul Cornu built the first manned helicopter in 1907.

■ The principle of the rotary wing was used some twenty-five hundred years ago by the Chinese for the flying top—a stick with propeller-like blades on top that was spun into the air as a children's toy.

■ More people are killed by donkeys every year than die in air crashes.

■ The hummingbird, the only bird capable of hovering in the air, beating its wings up to seventy times a second, provided the inspiration for the helicopter.

■ "Wet birds do not fly at night" was a code for the French Resistance during World War II.

WHY MONA LISA SMILES

In 1483, Leonardo da Vinci, inventor of the scissors, sketched a design for a helicopter.

pla/tic milk

WHAT YOU NEED

1 cup whole milk

Small saucepan

Measuring spoons

White vinegar

Clean, empty glass jar

Wax paper

WHAT TO DO

Pour the milk into the saucepan, add 2 teaspoons vinegar, and heat, stirring frequently. The milk will boil, then form tiny lumps (curds) in a clear liquid (whey). Slowly pour off the liquid from the pot into the sink. Then spoon the curds into the jar.

Add 1 teaspoon vinegar to the curds, and let the mixture stand for one to two hours.

The curds will form a yellowish glob at the bottom of a clear liquid. The glob is actually fat, minerals, and the protein casein. Pour out the liquid, remove the rubbery yellow glob from the jar, wash the glob with water, and knead it until it attains the consistency of dough.

Mold the plastic into any shape you wish, then place it on the wax paper. Let dry overnight.

WHAT HAPPENS

The casein from the milk hardens into plastic that can be painted with acrylic paints.

WHY IT WORKS

The combination of heat and acetic acid precipitates the casein, an ingredient used to make plastic, from the milk.

BIZARRE FACTS

■ The Japanese have developed a low-cost, biodegradable plastic made from shrimp shells by combining *chitin*—an extract from the shells that is also found in human fingernails—with silicon. The resulting "chitisand" is stronger than petroleum-based plastics, decomposes in soil, and acts as a fertilizer.

■ In 1929, the Borden Company purchased the Casein Company of America, the leading manufacturer of glues made from casein, a milk by-product. Borden introduced its first nonfood consumer product, Casco Glue, in 1932.

■ You can make glue from milk by simply adding ⅓ cup of vinegar to one cup of milk in a widemouthed jar. When the milk separates into curds and whey, pour off the liquid and wash it away. Add one-quarter cup of water and a tablespoon of baking soda. When the bubbling stops, you've got glue.

■ Cheese is made from curds. White glue is made from the casein of the curds.

■ While promoted by a spokesbull and made by a milk company, Elmer's Glue-All is a synthetic resin glue that does not contain casein.

■ Before scientists discovered how to synthesize plastics from petroleum products, plants and animal fats were used to make natural plastics, which eventually decompose.

■ Biodegradable plastic is made by adding starch to the plastic. Bacteria then feed on the buried plastic.

■ In surgery, stitches are now made using plastics that slowly dissolve in body fluids.

■ The American Dairy Council's ad campaign featuring celebrities with milk mustaches and the headline "Got Milk?" is translated in Mexico as "Are You Lactating?"

■ Twelve or more cows are called a *flink*.

■ The Sanskrit word for *war* means "desire for more cows."

■ There are more plastic flamingos in America than real ones.

■ In 1975, a Holstein cow in Indiana produced 195.5 pounds of milk in one day. That's enough to provide a hundred people with nearly a quart of milk each.

■ Cow's milk is 87 percent water.

- In Arctic regions, people get milk from reindeer.

- In Peru and Bolivia, people drink llama's milk.

- Twenty-four percent of all Americans drink milk with dinner.

- The glue on Israeli postage stamps is certified kosher.

- Ben and Jerry's sends the waste from making ice cream to local pig farmers to use as feed. Pigs love it, except for one flavor: Mint Oreo.

- On May 23, 1992, Ashrita Furman of Jamaica, New York, walked sixty-one miles with a full pint bottle of milk balanced on his head.

JUST ONE WORD

In the 1967 movie *The Graduate*, starring Dustin Hoffman, during a party in his parents' home to celebrate his college graduation, Benjamin Braddock is steered into the backyard by a concerned friend of his parents.

MR. McQUIRE: Ben, I want to say one word to you—just one word.

BENJAMIN: Yes, sir.

MR. McQUIRE: Are you listening?

BENJAMIN: Yes I am.

MR. McQUIRE: *(gravely)* Plastics.

BENJAMIN: What do you mean?

MR. McQUIRE: There is a great future in plastics. Think about it. Will you think about it?

BENJAMIN: Yes, I will.

MR. McQUIRE: Okay. Enough said. That's a deal.

pLAY douGH

WHAT YOU NEED

Large glass bowl

Food coloring

2 cups water

2 cups flour

1 cup salt

1 teaspoon alum

1 teaspoon 20 Mule Team Borax

1 tablespoon corn oil

Frying pan

Wooden spoon

Cutting board

Ziploc storage bag or airtight container

WHAT TO DO

In the bowl, combine the water and fifty drops of food coloring. Then add the flour, salt, alum, borax, and corn oil. Mix well. With adult supervision, cook and stir over medium heat for three minutes (or until the mixture holds together). Turn onto the cutting board and kneAD to proper consistency. Store in the Ziploc bag or airtight container.

WHAT HAPPENS

The alum and borax prevent bacteria from turning the thick, colored dough sour.

WHY IT WORKS

The patent for Play-Doh puts it this way: "We are unable to state definitely the theory upon which this process operates, because the reactions taking place in the mass are complicated."

BIZARRE FACTS

- In 1956 in Cincinnati, brothers Noah W. McVicker and Joseph S. McVicker, employees of Rainbow Crafts, a soap company, invented Play-Doh and received a patent for it in 1965. Kenner acquired Play-Doh, only to be bought out by Hasbro, which transferred Play-Doh to its Playskool division.

- The patent for Play-Doh (U.S. Patent No. 3,167,440) can be obtained by sending a written request and a check for three dollars to the Commissioner of Patents and Trademarks, Washington, D.C. 20321.

- After you make Play Dough in an old pan, the pan will be sparkling clean. The combination of flour and salt cleans the pan.

- Play-Doh is available in twenty-one colors.

- The Play-Doh boy, pictured on every can of Play-Doh, was created in 1960 and is named Play-Doh Pete.

MMMMM, GOOD!

Kids eat more Play-Doh than crayons, fingerpaint, and white paste *combined*.

quaker oats camera

WHAT YOU NEED

Empty, clean Quaker Oats cardboard canister with lid

X-acto knife

Paintbrush

Flat black tempera paint

Black electrical tape

Small piece of aluminum foil

#10 sewing needle

Wax paper

Scotch tape

Matches

Candle

Scissors

Black construction paper

Roll of 120 black-and-white film (slow speed)

Pitch-black room or closet

WHAT TO DO

Remove the lid of the canister. With adult supervision, use the X-acto knife to cut a hole about the size of a quarter in the center of the bottom of the canister. Paint the inside of the canister and the inside of the lid flat black. Let dry overnight.

Use the electrical tape to secure the piece of aluminum foil (shiny side facing in) over the hole so no light can get in. Use the needle to slowly drill a tiny, smooth, round hole in the aluminum foil, making sure the hole is centered in the bottom of the canister.

Cover the open end of the cardboard canister with one sheet of wax paper and fasten in place with the Scotch tape.

With adult supervision, light the candle in a dark room. Point the pinhole toward the candle. You will see an upside-down image of the candle on the wax paper.

Remove the wax paper from the canister.

Make a shutter by covering the pinhole of your Quaker Oats camera with a piece of the black construction paper. Tape one side with the Scotch tape so you can lift the paper like a flap.

Take the roll of film, the pinhole camera and lid, the scissors, the black electrical tape, and the Scotch tape into the dark room, placing them on a countertop where you can easily find them in the dark. Turn off the light.

Using your sense of touch alone, remove the roll of film from its canister and pull out about 8 inches of film from the roll. With the scissors, cut the film about an inch from the roll so you can get more film later. Cut about 3 inches from the starting end of the film (the end that sticks out of

the roll) and discard. This should leave you with about 4 inches of film. Place this strip centered inside the lid of the canister with the emulsion side facing up. Secure in place with Scotch tape at the two ends of the film. Replace the lid on the canister and tape it in place with the black electrical tape to keep out unwanted light. Replace the remainder of the film in its film canister. Hold the black construction paper shutter in place over the your hand over the pinhole and take your camera outside into bright sunlight.

Set the camera firmly on a solid surface to hold it steady. Aim it at a subject you would like to photograph. The subject must be still, with the sun shining on it.

When the camera is all set, fold back the flap, let the camera sit about one second to register a picture, then cover the pinhole with the flap again.

Return to your dark room or closet, remove the film from inside the lid of the Quaker Oats canister, place it in a film canister, and press the snap-on lid in place. Take your exposed film to a photo supply store, and have professionals develop the negative and make a print for you. (If your first picture turns out too dark, expose it longer the next time. If your picture is too light, expose it for a shorter time.)

WHAT HAPPENS

The homemade camera takes a picture just like an expensive state-of-the-art camera.

WHY IT WORKS

Light waves travel in a straight line from the subject to the pinhole. When the light waves go through the pinhole in the Quaker Oats canister, they project an upside-down image. The image is inverted because light travels in a straight line, causing the light from the top of the subject to strike the bottom part of the film. Light from the bottom of the subject strikes the top part of the film. The chemical emulsion on the film is sensitive to light and reacts to the image in front of the pinhole. When the film is developed a negative image is produced, which is then used to print the photograph.

BIZARRE FACTS

■ Light waves enter the lens of the eye and project an upside-down image on the retina, which our brains turn right side up again. If you were fitted with glasses that turned images upside down, after a while your brain would set things right for you, turning the images you see right side up again. When you took off the glasses, you would once again see the world upside down, until your brain readjusted and turned things right side up again.

■ In 1887, Henry D. Seymour, one of the founders of a new American oatmeal milling company, purportedly came across an article on the Quakers in an encyclopedia and was struck by the similarity between the religious group's qualities and the image he desired for oatmeal. A second story contends that Seymour's partner, William Heston, a descendant of Quakers, was walking in Cincinnati one day and saw a picture of William Penn, the English Quaker, and was similarly struck by the parallels in quality.

■ The name "Quaker Oats" inspired several lawsuits. The Quakers themselves unsuccessfully petitioned the United States Congress to bar trademarks with religious connotations.

■ Explorer Robert Peary carried Quaker Oats to the North Pole, and explorer Admiral Richard Byrd carried Quaker Oats to the South Pole.

- A gigantic likeness of the Quaker Oats man was placed on the White Cliffs of Dover in England, requiring an act of Parliament to have it removed.

- In 1990, when the Quaker Oats Company used Popeye the Sailor Man in oatmeal ads, the Society of Friends objected, insisting that Popeye's reliance on physical violence is incompatible with the religion's pacifist principles. The Quaker Oats Company quickly apologized and ended the campaign.

- About five hundred million pounds of aluminum foil are used in the United States every year. That's equal to eight million miles of aluminum foil.

SOWING YOUR OATS

In 1988, when nutritionists claimed that oat bran reduced cholesterol, sales for the Quaker Oats Company jumped 600 percent. In July 1992, a major report in *The Journal of the American Medical Association*, sponsored by the Quaker Oats Company, concluded that oat bran lowers blood cholesterol by an average of just 2 to 3 percent. On the bright side, the report claimed that a 1 percent reduction in cholesterol nationwide could lead to a 2 percent decrease in deaths from heart disease.

quicksand

WHAT YOU NEED

Newspaper

Measuring cup

1¼ cups cornstarch

1 cup water

Large bowl

Spoon

2 tablespoons ground coffee

WHAT TO DO

Cover the tabletop with newspaper. Combine the cornstarch and water in the bowl and stir until the mixture looks like a thick and sticky paste. Lightly sprinkle the ground coffee on top of the mixture to give it a dry, even look.

Make a fist and pound on the surface of the quicksand. Then lightly push your fingers down into the mixture.

WHAT HAPPENS

When you hit the surface of the mixture with your fist, the quicksand supports your weight, but when you push your fingers into the mixture, they easily sink to the bottom of the bowl. The coffee grounds make the mixture look deceptively smooth and dry, much like real quicksand.

WHY IT WORKS

The cornstarch-water mixture is a *hydrosol*—a solid dispersed in a liquid. When you punch the quicksand, your fist forces the long starch molecules closer together, trapping the water between the starch chains to form a semirigid structure. When the pressure is released, the cornstarch flows again.

BIZARRE FACTS

- Many people have lost their lives by sinking into quicksand—a thick body of sand grains mixed with water that appears to be a dry, hard surface. Although it looks as if it can be walked on, quicksand cannot support heavy weight.

- Cornstarch is the starch found in corn. All green plants manufacture starch through photosynthesis to serve as a metabolic reserve, but it wasn't until 1842 that Thomas Kingsford developed a technique for separating starch from corn, founding Kingsford's Corn Starch.

THINK QUICK

If you ever get caught in deep quicksand, fall flat on your back and stretch your arms out at right angles to your body—so you float on the quicksand. You can then slowly roll over onto firm ground.

- Cornstarch is an antidote for iodine poisoning.

- Cornstarch makes excellent spray starch for clothing. Mix one tablespoon of cornstarch and one pint of cold water. Stir to dissolve the cornstarch completely. Fill a spray bottle and use as you would any starch, making sure to shake vigorously before each use.

- Cornstarch makes an excellent substitute for baby powder and talcum powder. Cornstarch is actually more absorbent than talcum powder, but apply it lightly since it does cake more readily. (Talc, incidentally, is carcinogenic, particularly when inhaled or used by women in the genital region. Most baby powders contain talc).

- Cornstarch absorbs excess polish from furniture and cars. After polishing furniture or a car, sprinkle on a little cornstarch and rub with a soft cloth.

- Cornstarch has twice the thickening power of flour. When a gravy, sauce, soup, or stew recipe calls for flour, use half as much cornstarch to thicken. One tablespoon of cornstarch equals two tablespoons of flour.

VIVA QUICKSAND!

When Mexican revolutionary general Rodolfo Fierro, marching toward Sonora with Pancho Villa's troops in 1917, decided to take a shortcut, his horse got caught in quicksand. Fierro, loaded down with gold, sank to his death.

recycled paper

WHAT YOU NEED

Newspaper

Measuring cup

Clean, empty, large glass jar with lid

Hot tap water

Wooden spoon

Electric blender

3 tablespoons cornstarch

Metal baking pan (larger than 8 by 10 inches)

Metal screen (8 by 10 inches)

Scissors

Markers, crayons, paints, pencils, or pens

WHAT TO DO

Cut sheets of newspaper into long, thin strips (or feed the newspaper through a paper shred-der) until you have 1½ cups of packed, shredded newspaper.

Put the shredded newspaper into the jar and fill it three quarters full of hot tap water. Screw on the lid and let stand for three hours, shaking the jar occasionally and beating and stir-

ring with the wooden spoon. As the paper absorbs the water, add more hot tap water.

When the mixture becomes pasty and creamy, pour it into the blender—with adult supervision. Dissolve the cornstarch in $1/2$ cup hot tap water, pour into the blender, and blend. Pour the mixture into the baking pan.

Place the metal screen on top of the mixture in the baking pan, then gently push it down into the pan until the mixture covers it.

Bring the screen up, place it on a sheet of newspaper, and press it flat with the palm of your hand to squeeze away the water.

Let the screen-backed paper mixture dry in the sun for several hours. When the paper is thoroughly dry, peel it from the screen backing and trim the edges with scissors.

WHAT HAPPENS

The recycled newspaper has the texture of a gray egg carton and

can be decorated with markers, crayons, paints, pencils, or pens.

WHY IT WORKS

Newspaper pulp—a blend of sulfite pulp and ground cellulose fibers—when formed into a sheet over a screen, dries into paper again. The cornstarch—a sizing material—is added to the mixture to give the paper a smooth surface and prevent too much ink absorption. Any discarded paper—paper bags, computer punch cards, junk mail—can be made into pulp.

BIZARRE FACTS

- The first paper, invented in China in 105 C.E. by Ts'ai Lun, the Emperor Ho-Ti's minister of public works, was made from the inner bark of the mulberry tree, fishnets, old rags, and waste hemp.

- For hundreds of years, paper was made by hand from the pulp of rags. Rag pulp is still used today to make most high-quality bond paper.

- Toilet paper and facial tissues are made from wood pulp treated with plant resins to make it absorbent.

- The average American uses 640 pounds of paper and paperboard every year.

- Money is made out of cotton, not paper.

- Before the advent of paper, most documents were written on parchment (made from the skin of sheep or goats) or vellum (made from the skin of calves). A single book three hundred pages long would require the skins of an estimated eighteen sheep.

- The watermark was discovered by accident. In 1282, a small piece of wire caught in the paper press being used at the Fabrino Paper Mill made a line in the finished paper that could be seen by holding the paper up to the light. The papermakers realized a design made from wire would create a decorative watermark, which could also be used on banknotes to thwart counterfeiters.

PAPER TIGER

Paper can be made inexpensively from hemp. However, in 1937, cotton growers, fearing competition from hemp growers, lobbied against marijuana (the dried leaves of the hemp plant) to make hemp illegal. In 1999, Governor Jesse "The Body" Ventura signed legislation making hemp farming legal in Minnesota.

riſing golf bALL

WHAT YOU NEED

Golf ball

Clean, empty glass jar and lid

Uncooked rice

WHAT TO DO

Place the golf ball in the bottom of the jar, then fill the jar with the uncooked rice, stopping 1½ inches from the top of the jar. Close the lid. Shake the jar back and forth vigorously (not up and down).

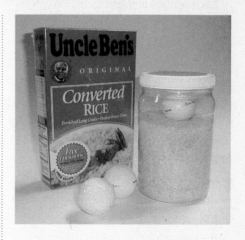

WHAT HAPPENS

The golf ball rises to the surface.

WHY IT WORKS

No two pieces of matter can occupy the same space at the same time. As the jar is shaken, the grains of rice move closer together, settling in the jar and pushing the ball upward.

BIZARRE FACTS

- More people play golf than any other outdoor sport.

- The nearly 11,700 golf courses in the United States occupy land valued at about $4.5 billion.

- In 1457, the parliament of King James II of Scotland banned golf and soccer because the popularity of the two sports threatened the practice of archery for national defense. The ban lasted until 1502, when England and Scotland signed a treaty of "perpetual peace."

- The "birdie"—the term for scoring one stroke under par on a hole—probably got its name from the "feathery"—the name for the original golf ball used until 1848, which was made from leather and stuffed with feathers.

- At the center of most golf balls is a sac filled with a liquid substance, usually castor oil and liquid silicone.

- The original Uncle Ben was a black rice farmer known to rice millers in and around Houston in the 1940s for consistently delivering the highest quality rice for milling. Uncle Ben harvested his rice with such care that he purportedly received several honors for full-kernel yields and quality. Legend holds that other rice growers proudly claimed their rice was "as good as Uncle Ben's." Unfortunately, further details of Uncle Ben's life (including his last name) were lost to history.

■ Frank Brown, a maître d' in a Houston restaurant, posed for the portrait of Uncle Ben.

■ In the 1980s, the portrait of Uncle Ben was dropped from the rice boxes for two years. Sales plummeted, and the company quickly reinstated Uncle Ben on the boxes.

■ Rice is thrown at weddings as a symbol of fertility.

■ The world's leading producer of rice is China. The world's second leading exporter of rice is Thailand, followed by the United States. The world's leading importer of rice is Iran.

■ Rice, grown on more than 10 percent of the earth's farmable surface, is the mainstay for nearly 40 percent of the world's population.

TOP SECRET RICE

During the processing necessary to produce white rice, the bran layer—containing a large part of the nutritive value of rice—is removed. In England, scientists discovered a special steeping and steaming process to force the bran nutrients, under pressure, into the rice grain *before* the bran is removed, locking the nutrients inside the grain.

In the early 1940s, George Harwell, a successful Texas food broker, received permission to introduce the process developed in England to the United States—but only if he could build a plant immediately. Because the new process improved the nutritional, cooking, and storage qualities of a food that had remained unchanged for more than five thousand years, Harwell convinced the United States government that this unique product merited war priorities. In 1943, Harwell and his partners shipped the first carload of Converted Brand Rice to an army quartermaster depot.

Until the end of World War II, Converted Brand Rice was produced for use solely by military personnel. Then, in 1946, Harwell's company, Converted Rice, Inc., brought this special rice to American consumers for the very first time using the familiar portrait of Uncle Ben as its trademark. Consumer response was so great that in just six years Uncle Ben's Converted Brand Rice became the number one packaged long grain rice sold in the United States.

rock candy

WHAT YOU NEED

Clean, empty glass jar

Measuring cup

Hot water

2 cups sugar

Spoon

Nail

String

Pencil

WHAT TO DO

Fill the jar with 1/4 cup hot tap water and slowly add the sugar, stirring well. Attach the nail to one end of the string and the pencil to the other end of the string so when you rest the pencil on the mouth of the jar, the nail hangs down into the thick sugar water without touching the bottom of the jar. Place the jar in a warm place and let it stand for a few days.

WHAT HAPPENS

The water evaporates and rocky sugar crystals form on the string.

WHY IT WORKS

As the water evaporates, the atoms in the sugar draw close together, forming cube-shaped crystals.

BIZARRE FACTS

- The average American uses ninety pounds of sugar every year.
- Sugar is used for mixing cement.
- In cooking, a small amount of sugar will make yeast work faster. Too much sugar will stop yeast from working at all.
- *Sugar* is slang for "money."

- During World War II, GIs called a letter from one's sweetheart a "sugar report."
- Addressing the Canadian Senate and House of Commons in 1941, Winston Churchill said, "We have not journeyed all this way across the centuries, across the oceans, across the mountains, across the prairies, because we are made of sugar candy."
- Adding a tablespoon of sugar per quart of water in a vase prolongs the life of fresh flowers.
- Sprinkling a dash of sugar on a tongue burned by hot soup, tea, or coffee soothes the burn.

SUGAR SUGAR

The hit song "Sugar Sugar," by the Archies, was the number three bestselling single in 1969, topping the Rolling Stones' "Honky Tonk Woman" and the Beatles' "Get Back."

rubber chicken bone

WHAT YOU NEED

Uncooked chicken bone or
wishbone

Clean, empty glass jar with lid

White vinegar

WHAT TO DO

Clean the bone thoroughly and let dry overnight. Place the bone in the jar and add enough vinegar to cover the bone. Secure the lid and let the jar stand undisturbed for seven days.

WHAT HAPPENS

The bone becomes soft and rubbery. It can be twisted, and, in some cases, tied in a knot.

WHY IT WORKS

The vinegar (acetic acid) dissolves the calcium from the bone, leaving it soft and bendable.

BIZARRE FACTS

■ The Etruscans (the people of ancient Northern Italy) originated the superstition of having two people each make a wish and tug on the opposite ends of the dried V-shaped clavicle of a fowl. The person who breaks off the larger piece of the "wishbone" allegedly has his or her wish come true. Etymologists claim this custom gave birth to the expression "get a lucky break."

■ *Boneyard* is slang for "cemetery."

■ Before double murderer Robert Harris was executed in California in 1992, his last meal included a bucket of Kentucky Fried Chicken.

■ Colonel Sanders told the *New York Times* that the eleven herbs and spices in Kentucky Fried Chicken "stand on everybody's shelf." In reality, there are only four, according to author William Poundstone, none of which are herbs: flour, salt, monosodium glutamate, and black pepper.

THE FUNKY CHICKEN

Colonel Harland Sanders, who, at the age of seventy, founded Kentucky Fried

Chicken, talked his wife into hiring his mistress as their live-in housekeeper, according to the Colonel's daughter, Margaret Sanders, in her book, *The Colonel's Secret: Eleven Secret Herbs and a Spicy Daughter.* Sanders later divorced his wife, married his mistress, and took both women with him to Washington, D.C., to attend a presidential inauguration.

rubber egg

WHAT YOU NEED
Hard-boiled egg
Clean, empty glass jar and lid
White vinegar

WHAT TO DO
Place the egg in the jar. Cover with vinegar. Secure the lid and let the jar stand for one week.

WHAT HAPPENS
The eggshell dissolves and you are left with a rubbery egg that will actually bounce if not dropped from too great a height.

WHY IT WORKS
The vinegar (acetic acid) dissolves the calcium carbonate in the eggshell.

BIZARRE FACTS

■ English author Samuel Butler wrote: "A hen is only an egg's way of making another egg."

■ As many as nine yolks have been found in hen's eggs.

■ On February 2, 1990, at the International Poultry Show in Atlanta, Georgia, Howard Helmer made 427 two-egg omelettes in a half hour, setting a world record.

■ On April 9, 1992, the staff of Cadbury Red Tulip in Ringwood, Australia, completed the heaviest and tallest Easter egg on record, weighing 10,482 pounds 14 ounces and reaching a height of 23 feet 3 inches.

■ The working title of the Beatles' song "Yesterday" was "Scrambled Eggs."

■ The average ostrich egg, approximately twenty-four times the size of a hen's egg, can support the weight of a 280-pound human.

■ The vervain hummingbird of Jamaica lays the smallest bird egg in the world, weighing less than 0.0132 ounce.

HOLY EGG SHELLS!

On the 1960s television show *Batman*, the arch-criminal Egghead, played by Vincent Price, was considered the smartest villain in the world, hatched fiendish schemes, and had an *eggs-cellent* vocabulary, including words like *eggs-traordinary, eggs-citing,* and *eggs-hilirating.*

Secret Message Egg

WHAT YOU NEED
Alum
White vinegar
Q-tips cotton swabs
Hard-boiled egg
Small bowl
Measuring cup

WHAT TO DO

In the bowl, dissolve one part alum in one part vinegar. Mix well. Use the Q-tips cotton swab to write or draw something on the eggshell using the alum-and-vinegar solution as ink. Let dry. Remove the eggshell from the egg.

WHAT HAPPENS

The alum-and-vinegar solution dries invisible, but when the eggshell is removed, the writing is visible on the egg's surface (and the egg is still edible).

WHY IT WORKS

The vinegar (acetic acid) dissolves the calcium carbonate in the eggshell, allowing the alum to permeate the shell and discolor the egg white.

BIZARRE FACTS

■ In parts of Germany during the 1880s, Easter eggs etched with the recipient's name and birth date were accepted in courts of law as birth certificates.

■ In the 1880s, Russian czar Alexander III commissioned goldsmith Peter Carl Fabergé to craft Easter eggs for his wife, czarina Maria Feodorovna. The Soviet government took over Fabergé's business after the Bolshevik Revolution of 1917. Forty-three of the original fifty-three eggs known to have been made by Fabergé are in museums and private collections and are collectively valued at over $4 million.

■ In *St. Ives*, Robert Louis Stevenson wrote, "You cannot make an omelette without breaking eggs," inspired by the French proverb, "*On ne fait pas d'omelette sans casser des oeufs.*"

■ In *Pudd'nhead Wilson*, Mark Twain wrote: "Put all your eggs in one basket and—*watch that basket.*"

- On September 6, 1981, in Siilinjärvi, Finland, Risto Arntikainen threw a fresh hen's egg 317 feet 10 inches to Jyrki Korhonen, without breaking it.

- According to the Food and Agriculture Organization of the United Nations, as of 1985 there were 8,295,760,000 chickens on the planet. That's 1.6 chickens for every person in the world.

- In 1990, British jeweler Paul Kutchinsky unveiled a two-foot-tall jeweled egg made from thirty-seven pounds of gold and stuffed with twenty thousand pink diamonds, now on display in London's Victoria and Albert Museum.

THE HEN THAT LAID THE GOLDEN EGG

In 1979, the College of Agriculture at the University of Missouri recorded a hen that laid 371 eggs in 364 days.

ſeeſawing candle

WHAT YOU NEED
Knife

Candle

Needle

Wax paper

Two drinking glasses

Matches

WHAT TO DO

With adult supervision, carve away the tallow or wax at the bottom end of the candle to expose the wick. Carefully push the needle through the center of the candle.

Place a piece of wax paper on the tabletop, set the drinking glasses on the wax paper, and rest the needle across the rims of each glass so the candle is between the glasses.

Light both ends of the candle, and give one end a slight push so the candle teeters like a seesaw.

WHAT HAPPENS

The candle starts rocking like a seesaw and continues as long as both ends stay lit.

WHY IT WORKS

As Newton's third law states, "For every action there is an opposite and equal reaction." When the tallow or wax drips off each end of the candle, it delivers a slight upward recoil.

BIZARRE FACTS

■ During Roman times, starving soldiers ate their candle rations, which were made from tallow—a colorless, tasteless extract of animal or vegetable fat.

■ British lighthouse keepers, isolated for months, often ate their tallow candles.

■ The charred end of the wick of a tallow candle had to be "snuffed"— snipped off without extinguishing the flame—every half hour, otherwise the candle provided only a fraction of its potential light and the low-burning flame burned only 5 percent of the tallow, melting the remaining tallow.

■ Because snuffing a tallow candle was difficult to do without extinguishing the flame, the word *snuff* came to mean "extinguish."

- Beeswax oozes from small pores in the abdomen of a worker bee and forms tiny white flakes on the outside of its abdomen. Using its legs, the bee picks off these flakes and moves them to its jaws. The bee then chews the wax to the proper consistency to build the honeycomb.

- Native Americans in the Pacific Northwest used dried candlefish, a saltwater fish about eight inches long, as candles.

- Spermaceti, a waxy material obtained from the enormous head of the sperm whale (making up a third of its body), was once used to make candles.

- The world's largest candle on record, displayed at the 1897 Stockholm Exhibition, was eighty feet high and 8$\frac{1}{2}$ feet in diameter.

- The world's largest needle, measuring 6 feet 1 inch long, used for stitching on mattress buttons lengthwise, can be visited at the National Needle Museum in Forge Mill, Great Britain.

BURN, BABY, BURN!

My candle burns at both ends;
It will not last the night;
But, ah, my foes, and oh, my friends—
It gives a lovely light.

—EDNA ST. VINCENT MILLAY

ſmoke bomb

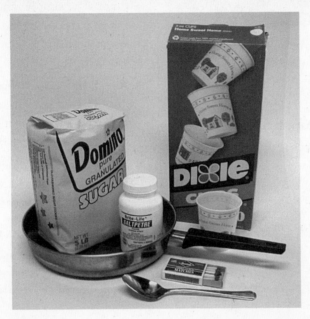

WHAT YOU NEED

¼ cup sugar

3 ounces saltpetre (from a drugstore or medical supply store)

Bowl

Saucepan

Spoon

Paper cup

Wooden matches

WHAT TO DO

With adult supervision, combine the sugar and saltpetre in the bowl, then heat in the saucepan over very low heat, stirring constantly, until the mixture melts into a plastic substance that resembles caramel. Remove from the heat, pour into the paper cup, embed a few wooden matches, heads up, into the hardening substance, and let cool.

When the smoke bomb cools and hardens, peel off the paper cup, place the smoke bomb inside a clean, empty tin can or on a concrete surface (away from flammable objects or areas), light the match heads, and stand back. (Or you can simply leave

the paper cup in place and light it on fire.)

WHAT HAPPENS

The smoke bomb produces enough thick white smoke to fill a small room.

WHY IT WORKS

The melted sugar becomes a candy that encapsulates the saltpetre, making it easier to ignite.

BIZARRE FACTS

- In ancient Chinese kitchens, saltpetre was commonly used as a preserving and pickling salt.

- Saltpetre is one of the three original ingredients in fireworks.

- Smoke is simply small particles made airborne by the buoyancy of the hot gases produced by combustion.

- Along the Great Wall of China, guards sent smoke signals made with wolf dung, because the smoke hung in the air for a long time.

- In the movie *October Sky*, the rocket boys mix saltpetre and sugar to make rocket fuel.

THE JOYS OF SALTPETRE

Saltpetre, known to chemists as potassium nitrate, is commonly believed to reduce a man's primitive urges. It doesn't.

spraying fountain

WHAT YOU NEED

Electric drill with ¼-inch bit

Two clean, empty glass jars and one plastic lid

Two straws

Caulk

Water

Yellow food coloring

Blue food coloring

Baking pan

WHAT TO DO

With adult supervision, drill two holes in the lid. Insert the first straw through a hole so that 2 inches of straw extend above the lid. Insert the second straw through the other hole in the lid so that 2 inches extend inside the lid. Caulk the holes around the straw and the lid.

Fill one jar half full with water, add five drops of yellow food coloring, stir well, and screw on the lid. Fill the second jar with water, add five drops of blue food coloring, and stir well.

Place the jar of blue water in the baking pan. Turn the jar with the lid and straws upside down, and place the shorter straw in the jar, letting the

longer straw empty yellow water into the pan.

WHAT HAPPENS

As the yellow water empties into the baking pan, the blue water rises and sprays like a fountain in the sealed jar.

WHY IT WORKS

Gravity empties the yellow water from the sealed jar through the straw, reducing the air pressure inside the jar. The air pressure outside the sealed jar, now greater than the air pressure inside the jar, pushes down on the blue water in the second jar, forcing it to spray out of the other straw.

BIZARRE FACTS

■ While pumps supply the pressure in artificially created fountains, the enormous weight of water in a reservoir generates the pressure for natural fountains.

■ The fountain at Fountain Hills, Arizona, is the tallest fountain in the world, creating a column of water up to 625 feet tall and weighing more than eight tons.

■ According to ancient Greek legend, drinking the water from the fountain of Castalia on the sacred mountain Parnassus bestows the ability to write poetry.

■ The 1954 Academy Award–winning movie *Three Coins in a Fountain*, following the stories of three women who each toss a coin into the Trevi Fountain in Rome, was remade in Madrid as the 1964 movie *The Pleasure Seekers*.

FOUNTAIN OF YOUTH

Spanish explorer Ponce de León searched Florida for the mythical Fountain of Youth, a spring whose waters would reputedly make old people young and heal the sick. He did find a spring in St. Augustine that he thought would give him eternal youth, and today you can visit the Fountain of Youth at 155 Magnolia Street and admire a statue of Ponce de León that does not age.

ſteel wool ſparkler

WHAT YOU NEED

Steel wool pad

D battery

1 baking pan

WHAT TO DO

With adult supervision, pull the steel wool pad apart until it is the size of a tennis ball. Place the steel wool pad in the baking pan. Touch the ends of the battery to the steel wool.

WHAT HAPPENS

The sparks from the battery cause the steel wool to catch on fire, and the iron filings from the steel wool sparkle like a Fourth of July sparkler.

WHY IT WORKS

The threads of iron in the steel wool, surrounded by more oxygen than in a solid block of iron, combust easily.

BIZARRE FACTS

■ In 1917, Edwin Cox, a struggling door-to-door aluminum cookware salesman in San Francisco, developed in his kitchen a steel wool scouring pad caked with dried soap as a gift to housewives to get himself invited inside their homes to demonstrate his wares and boost sales. A few months later, demand for the soap-encrusted pads snowballed, and Cox quit the aluminum cookware business and went to work for himself.

■ Mrs. Edwin Cox, the inventor's wife, named the soap pads S.O.S., for "Save Our Saucepans," convinced that she had cleverly adapted the Morse code international distress signal for "Save Our Ships." In fact, the distress signal S.O.S. doesn't stand for anything. It's simply a combination of three letters represented by three identical marks (the S is three dots, the O is three dashes). The period after the last S was deleted from the brand name to obtain a trademark for what would otherwise be an international distress symbol.

■ *Brillo* is a derogatory name for a person with tight curly hair.

....

BE PREPARED

The Boy Scout manual now instructs all Boy Scouts to start a fire with a steel wool pad and a D battery rather than the traditional method of starting a fire with flint and steel.

stink bombs

AMMONIUM SULFIDE STINK BOMB

WHAT YOU NEED

Single-edge razor blade

Box of wooden safety matches

Clean, empty glass jar with lid

1 cup ammonia

WHAT TO DO

With adult supervision, use the razor blade to carefully cut off the heads of all the matches. Place the match heads in the jar. Add the ammonia to the jar and immediately screw on the lid tightly. Let it sit for one week. The solution can be thrown or poured directly in an area you think already stinks. Or place the jar uncapped behind a door that opens into the target room. When someone enters, the door will knock over the jar, spilling the liquid.

WHAT HAPPENS

The solution stinks like rotten eggs.

WHY IT WORKS

The sulfur in the match heads combines with the ammonia to form ammonium sulfide.

ROTTEN EGG STINK BOMB

WHAT YOU NEED

Crystal Drano (sodium hydroxide)

Clean, empty glass jar with lid

Warm water

Six egg whites

Spoon

WHAT TO DO

With adult supervision, pour about ¼ to ½ inch Crystal Drano into the jar, along with about an inch of warm water and the egg whites. Stir well. Screw the lid on the jar and let the mixture sit in a safe, warm place for one week. Shake well. The solution can be thrown or poured directly in an area you think already stinks. Or place the jar uncapped behind a door that opens into the target room. When someone enters, the door will knock over the jar, spilling the liquid. Make sure you don't get any on yourself.

WHAT HAPPENS

The solution emits a vile aroma.

WHY IT WORKS

The sodium hydroxide rots the egg whites.

BIZARRE FACTS

■ The origin of the children's refrain, "Last one there is a rotten egg," is unknown.

■ In 1830, French inventor Charles Suria produced a reliable match with a tip coated with white phosphorus—a chemical that, unfortunately, gives off poisonous fumes which cause "phossy jaw," a fatal disease that rots away the jaw-bone. By 1900, white phosphorus was banned.

■ In 1871, when the British Parliament considered imposing a "match tax" at a penny a box, thousands of match workers protested. The resulting riots prompted Parliament to strike the tax.

■ The most vile-smelling substances in the world, according to *The Guinness Book of Records*, are ethyl mercaptan and butyl seleno-mercaptan, "each with a smell reminiscent of a combination of rotting cabbage, garlic, onions, burned toast, and sewer gas."

■ The skunk sprays its foul-smelling musk up to twelve feet. The odor remains for several days.

■ In nearly every Pepe LePew cartoon, Pepe the skunk falls in love with a black cat that accidentally gets a white stripe painted down its back.

■ The skunk cabbage, a plant that grows in low swamps in eastern and central North America, is named for its unpleasant odor.

■ John Water's 1981 movie *Polyester*, starring Divine and Tab Hunter, was filmed in "Odorama." Viewers experienced the movie's various smells—including vomit and excrement—with a scratch-and-sniff card.

- Receptor cells in the sinus cavity of the nose send the impulses created by odors along the olfactory nerves to the olfactory bulb in the brain, which then sends the nerve impulses to the forebrain. The human olfactory bulb is small, while the olfactory bulb is large in dogs and other animals more reliant on the sense of smell.

- A dog possesses a sense of smell that is 1 million to 100 million times keener than a human's.

- Scientists do not know exactly how the brain distinguishes different smells.

- Smelling salts are made with ammonia.

- In humans, the sense of smell grows accustomed to a strong odor after prolonged exposure and automatically turns itself down.

C'MON, GET HAPPY

In an episode of *The Partridge Family,* a skunk stows away aboard the Partridge Family's bus. Mr. Kincaid forces America's grooviest television family to bathe in tomato juice to get rid of the smell.

ſwinging cupſ

WHAT YOU NEED

Two chairs

Thin rope

Ruler

Scissors

Two coffee cups

WHAT TO DO

Place the chairs back to back 3 feet apart. Tie a piece of the rope from the top of one chair to the top of the other.

Measure and cut two pieces of thin rope 24 inches long and tie a coffee cup to the end of each rope. Tie the free ends of the ropes to the rope between the chairs, letting the coffee cups hang 20 inches apart from each other and equidistant from the chairs. Move the chairs so that the center of the horizontal rope sags about 4 inches below the spots where the cord is tied to the chair backs.

Start one of the cups swinging at right angles to the horizontal cord like a pendulum.

Tell your audience to watch closely. Insist that by using your hypnotic energy, you will make the first cup stop swinging and the second cup start swinging. Wiggle your fingers toward the cups. The first cup slows down until it stops completely while the second cup begins swinging. A moment later, wiggle your fingers toward the cups again. The second cup will stop swinging and the first one will begin swinging again.

WHAT HAPPENS

While you pretend to be controlling the cups hypnotically, the motion continues being transferred from one cup to the other as long as the cups keep swinging.

WHY IT WORKS

Pendulums attached to a central line transfer energy back and forth between themselves through the connecting line.

BIZARRE FACTS

- Galileo Galilei first demonstrated this effect in 1583.

- At age twenty, Galileo discovered the law of the pendulum by timing the swings of a bronze lamp that hangs from the ceiling in the cathedral in Pisa, Italy. He observed that each swing took the same time, whether the arc was large or small.

- Edgar Allan Poe, author of *The Pit and the Pendulum*, married his cousin Virginia Clemm when she was not quite fourteen years old.

WORLD'S LARGEST YO-YO

On March 29, 1990, the woodworking class of Shakamak High School in Jasonville, Indiana, launched a yo-yo measuring six feet in diameter and weighing 820 pounds from a 160-foot crane. It "yo-yoed" twelve times. The word yo-yo means "come-come" in Filipino, and the toy purportedly originated in the Philippines as a weapon.

ʃwitching ʃwitcheʃ

WHAT YOU NEED

Lightbulb base

Plastic project enclosure box (6 by 3 by 2 inches)

Electric drill with ⅜-inch bit and ¹⁄₁₆-inch bit

Precision screwdrivers

Needle-nose pliers

Three push-on/push-off switches

Sharpie marker

Krazy Glue

AA battery holder

Wire cutters

Wire

Soldering iron

Solder

AA battery

2-volt lightbulb

WHAT TO DO

Mount the lightbulb base on the cover panel of the project enclosure box. With adult supervision, use the drill with the ⅜-inch bit to mount the three push-on/push-off switches under the lightbulb base. Use the Sharpie marker to label them A, B, and C. Using Krazy Glue, adhere the battery holder to the underside of the cover panel. Wire the switches in series (as illustrated in the schematic diagram) to the battery holder and lightbulb base. With an adult watching, solder the connections. Insert the battery and lightbulb. Screw the cover panel into place on the project enclosure box.

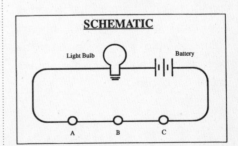

Press all the buttons so the bulb lights. Push button A several times to show that it controls the bulb. Leave the switch off. Push button B an odd number of times to show that it does not affect the bulb. Then push button C an odd number of times to show that it does not affect the bulb. This leaves all the buttons turned off.

Ask a volunteer to push whatever button he thinks will turn on the bulb. He naturally pushes button A, but the bulb does not light. Give him a second chance. He pushes button B, but again the bulb does not light. You push button C, which miraculously lights the bulb.

Repeat the trick by pushing button C several times to show it does indeed light the bulb, then leave it off. Push the other two buttons an odd number of times to show that they do not light the bulb. Once again, ask the volunteer to push whatever button he thinks will turn on the bulb.

WHAT HAPPENS

No matter how many times you repeat the trick, the volunteer continues failing on his first two tries. The third button selected always lights the bulb. Most people assume there is a perplexing circuit pattern under the board.

WHY IT WORKS

The switches and the bulb are joined in series to the battery.

All the switches are turned off when the volunteer is asked to push a button. His first choice turns on the first button, his second choice turns on the second button, and the third button (which you push) completes the circuit and lights the bulb.

BIZARRE FACTS

■ The trick of making a pea vanish under one of three thimbles originated in Egypt some five thousand years ago using cups and balls. The trick is done with slight of hand. The conjurer lets the pea roll out from under the cup, quickly catches it in the hand holding the cup, and then drops it into another thimble.

■ Three Card Monte—a card trick in which a gullible spectator is asked to wager money on which one of the three cards lying face down on a table is the queen—is rigged by a fake throw. The operator, previously throwing the bottom card from his right hand before the top card, unexpectedly throws the top card from his right hand before the bottom one.

tornado machine

WHAT YOU NEED

Sandpaper

Two clean, empty 2-liter soda bottles with plastic screw-on caps

Krazy Glue

Black electrical tape

Electric drill with ⅜-inch bit

Water

Blue food coloring

Measuring spoons

Silver glitter

WHAT TO DO

Sand the tops of the two screw-on bottle caps, then glue the tops of the caps together with Krazy Glue. Let dry. Wrap black electrical tape around the circumference of the two caps to secure them together. With adult supervision, drill a ⅜-inch hole through the center of the two caps.

Fill one of the bottles two-thirds full with water; add five drops of blue food coloring and 1 tablespoon silver glitter. Thread the joined caps onto this bottle, then thread the second bottle to the free end of the cap.

Turn the Tornado Machine upside down so the blue glitter water pours into the empty bottle. Swirl the full bottle counterclockwise until a tornado funnel forms.

WHAT HAPPENS

The water whirls down into the empty bottle like a tornado.

WHY IT WORKS

The swirling water spins faster toward the hole, creating a vortex as the water molecules come closer to the center. The resulting outward force pushes the liquid out of the center, creating a funnel.

BIZARRE FACTS

■ According to a study of 304 tornadoes in the United States between

1950 and 1991, tornadoes are more likely to strike on May 16 than any other day of the year.

■ The winds of a tornado whirl counterclockwise north of the equator and clockwise south of the equator.

■ The average tornado lasts less than thirty minutes and travels about twenty miles at ten to twenty-five miles per hour.

■ The overwhelming majority of tornadoes—approximately seven hundred a year—occur in the central and southeast United States, better known as Tornado Alley.

■ The calm center of a tornado as it passes overhead lasts from two minutes to a half hour.

■ When a tornado touches the surface of an ocean or lake, it becomes a waterspout, sucking water up inside the spinning wind. Waterspouts appear to rise up out of the water like a sea monster, possibly explaining the origin of those legends.

I'LL GET YOU, MY PRETTY

The tornado in *The Wizard of Oz* was actually a funnel made of muslin stiffened with wire. To bring the muslin tornado to life, a prop man was lowered inside the muslin tube to pull the wires in and out.

underwater candle

WHAT YOU NEED

Utility knife

Corrugated cardboard box (roughly 1½ by 2 feet wide and 1½ feet high)

Piece of corrugated cardboard (roughly 1½ by 2½ feet)

Ruler

Pencil

Newspaper

Paintbrush

Flat black tempéra paint

Clear packaging tape

Glass from a picture frame (about 8 by 10 inches)

Candle

Drinking glass

Matches

Pitcher

Water

WHAT TO DO

With adult supervision, use the utility knife to carefully cut the flaps from the top of the box, and fit the long piece of cardboard snugly upright diagonally inside the box (from the front left-hand corner to the back right-hand corner), dividing the box equally into two triangular halves.

Carefully measure and cut a small viewing window (6 inches square) in the front left half of the box where the dividing cardboard meets the corner.

Remove the divider from the box and cut a window about 7 by 9 inches that will line up with the window on the box, allowing you to look straight through to the left rear side of the box.

Spread newspaper on the tabletop and paint the divider and the inside of the box with the flat black tempera paint. When the paint is dry,

use the packaging tape to secure the 8-by-10-inch glass over the opening in the divider. Place the divider back in the box.

Mount the candle in the center of the first compartment. Place the drinking glass in the center of the rear compartment. Light the candle. Look through the viewing window and through the glass window. Align the candle and the glass until the candle appears in the center of the glass.

Have a friend look through the viewing window. Slowly pour the pitcher of water into the empty glass in the rear compartment.

WHAT HAPPENS

Your friend sees a candle burning in a glass of water.

WHY IT WORKS

The light reflecting from the front of the glass window reflects the image of the candle in the glass. This optical illusion is created because you can still see through the window to the glass with the water.

BIZARRE FACTS

■ This optical illusion is used in circus sideshows to create the illusion of a human turning into a skeleton or monster.

■ In 1862, after fifteen years of work, French physicist Jean Foucault used a candle and two mirrors to measure the speed of light at 187,000 miles per second. Modern methods have refined the figure to 186,282 miles per second.

■ Our eyes remain the same size from birth, but our nose and ears never stop growing.

■ An ostrich's eye is bigger than its brain.

IT'S A BIRD, IT'S A PLANE

In comic books, when Superman uses his X-ray vision to see through concrete walls, X rays shoot from his eyes. Living creatures do not see by radiating light from their eyes. (X rays are a more energetic form of light.) Instead, light enters the eye, is refracted by the cornea, passes through the pupil, is refracted by the lens, and forms an image on the retina, which is rich in nerve cells that are stimulated by light. To see through concrete walls, the nerve cells in Superman's retinas would have to be stimulated by X rays.

underwater fireworks

WHAT YOU NEED

Large, clear glass bowl

Water

1 tablespoon vegetable oil

Paper cup

Food coloring—red, blue, green

Spoon

WHAT TO DO

Fill the bowl with water. Pour the oil into the paper cup. Add four drops of each food coloring color. Mix the oil and colors thoroughly with the spoon. Pour the colored oil mixture into the water in the bowl. Observe for ten minutes.

WHAT HAPPENS

Small pools of oil spotted with tiny spheres of color float to the sur-face of the water, exploding outward and creating flat circles of color on the surface of the water. Long stream-ers of color then sink down through the water, like a fireworks display.

WHY IT WORKS

Oil and water are *immiscible*—meaning they do not mix, but separate into layers because of the different polarity of their molecules. The oil rises to the surface because it is less dense than the water. Since the water-based food coloring does not dissolve in oil, it remains in tiny spheres throughout the oil on the water's surface, then sinks through the oil layer and dissolves in the water below, creating long streamers of color.

BIZARRE FACTS

- In the tenth century, the Chinese discovered that three common kitchen ingredients—saltpetre, sulfur, and charcoal—were explosive when combined. When packed into a bamboo tube and ignited, the mixture rocketed skyward and exploded, lighting up the sky.

- In fireworks, sodium compounds produce yellow light, strontium and lithium salts emit red light, copper gives blue light, and barium creates green light.

- In 1988, the world's longest fireworks display—measuring 18,777 feet long, consisting of 3,338,777 firecrackers and 1,468 pounds of gunpowder—was ignited in Johor, Malaysia, and burned for nine hours and twenty-seven minutes.

- On July 15, 1988, the world's largest firework—weighing 1,543 pounds and measuring 1,354.7 inches in diameter—was exploded over Hokkaido, Japan, bursting to a diameter of 3,937 feet.

FLYING TOAST

Chinese inventor Wan-hu attempted to fly by building a plane made from two kites, a chair, and forty-two rocketlike fireworks. When the firecrackers were ignited, Wan-hu and his contraption went up in flames.

VOLCANO AT THE BEACH

WHAT YOU NEED

Clean, empty liter soda bottle

1 cup water

1 tablespoon baking soda

1 tablespoon liquid dishwashing detergent

Red food coloring

Plastic dinosaurs or army soldiers

1 cup white vinegar

WHAT TO DO

Fill the bottle with the water, baking soda, liquid dishwashing detergent, and ten drops of red food coloring. Place the bottle on the beach and build a sand volcano around the bottle. Place the plastic dinosaurs or soldiers around the volcano. Pour the vinegar into the bottle and scream "Volcano! Volcano! Volcano!"

WHAT HAPPENS

Red foam "lava" will bubble and spray out the top of the volcano and down the mountain of sand, covering the dinosaurs or soldiers.

WHY IT WORKS

The baking soda (a base) reacts with the vinegar (an acid) to produce carbon dioxide, producing foam and forcing the liquid out of the bottle.

VOLCANIC BLUNDER

The 1968 motion picture *Krakatoa, East of Java*, starring Maximilian Schell, Brian Keith, and Sal Mineo, follows the SS *Batavia Queen* as it is engulfed by the huge tidal wave created by the 1883 eruption of the volcanic island of Krakatoa. The eruption was heard about three thousand miles away, created sea waves almost 130 feet high, and killed nearly thirty-six thousand people on nearby islands. Oddly, the filmmakers mistitled their movie. Krakatoa lies west of Java. It is east of Sumatra.

EXPLODING VOLCANO

WHAT YOU NEED

Scissors

Metal screening

Needle

Thread

Newspaper

Circle of ¼-inch thick pine wood, 1 foot in diameter

Hammer

Carpet tacks (or staple gun and staples)

Clean, 3-ounce cat food can

Measuring cup

Plaster of Paris

Water

Bowl

Paintbrush

Brown paint

Shellac

Ammonium dichromate (from a chemical supply store)

WHAT TO DO

With adult supervision, and using scissors, form a cone shape with the metal screening, and sew it together with the needle and thread. Fill the cone with crumpled-up newspaper, and tack it down to the pine board. Set the cat food can in the cone of the volcano.

Cover the tabletop with newspaper. Cut newspaper strips 1 to 2 inches in width. Mix plaster of Paris and water in a bowl, according to the package instructions. Dip each newspaper strip into the plaster, gently pull it between your fingers to remove excess plaster, and apply it to the screening until it is completely covered. Let it dry, then paint it with the brown paint. Spray with shellac to seal the volcano. Let dry thoroughly.

Set the volcano outdoors. Crumple up a small piece of newspaper, put it in the cat food can, then pour some ammonium dichromate over it. With an adult watching, light the newspaper with a match, then quickly back away. (Instead of ammonium dichromate, you can place a smoke bomb from page 95 in the volcano and ignite it.)

WHAT HAPPENS

The volcano flares up, sparks, sputters, and sends a green lava flow spilling over the sides.

WHY IT WORKS

When ignited, ammonium dichromate (an orange, crystalline solid) spews sparks, a fluffy green solid (chromium oxide), steam (water vapor), and heat (nitrogen gas).

THE VOLCANO FARM

On February 20, 1943, a volcano began forming from a crack in the earth in a farmer's cornfield in Paricutín, Mexico (near the southwestern city of Uruapan). Within six days, the volcanic material formed a cinder cone over five hundred feet high. Two months later, the cone reached one thousand feet. The lava destroyed the village of Paricutín and San Juan Paragaricútiru. Today the volcano, which ceased activity in 1952, stands 1,345 feet high.

UNDERWATER VOLCANO

WHAT YOU NEED

Large pot

Cold water

Ice cubes

Spoon

Clean, empty baby food jar

Hot tap water

Red food coloring

Marbles

WHAT TO DO

Fill the pot with cold water and put in some ice cubes. When the water is sufficiently cold, fish out the ice cubes with the spoon. Fill the baby food jar three-quarters full with hot tap water and add five drops of red food coloring. Stir well. Add five or six marbles to give the jar some weight. Place the jar in the bottom of the pot.

WHAT HAPPENS

The red water slowly "erupts" from the jar toward the surface of the water. When the red water cools, it settles at the bottom of the pot.

WHY IT WORKS

Heat rises. Hot water is lighter than cold water because the molecules in hot water are farther apart than the molecules in cold water.

BIZARRE FACTS

■ Pumice, a natural glass that comes from lava, is widely used to remove calluses.

■ In Reykjavik, Iceland, most people heat their homes with water piped from volcanic hot springs.

■ When Mount Tambora in Indonesia erupted in 1815, it released six million times more energy than an atomic bomb and killed over twelve thousand people.

■ In 1963, an underwater eruption began forming the island of Surtsey in the North Atlantic Ocean. After the last eruption of lava in 1967, the island covered more than one square mile.

■ History's most famous volcanic eruption, Mount Vesuvius, took place in 79 C.E., and destroyed the Italian towns of Herculaneum, Pompeii, and Stabiae. Pompeii remained untouched beneath the ash deposits for almost seventeen hundred years.

■ The word volcano comes from Vulcan, the Roman god of fire who lived beneath a volcanic island called Vulcano off the coast of Italy.

■ The Voyager mission's photograph of Io, Jupiter's third-largest moon, showed ten active volcanoes, some of them erupting—making Io more volcanically active than Earth.

MIRACLE ON BALI

Besakih Temple, the holiest temple on the island of Bali, sits at the foot of the Gunung Agung volcano. On this holy spot, the Eka Dasa Rudra festival is held once every one hundred years—meaning that all the traditions needed to prepare for the festival must be passed down from one generation to the next. On the last day of the celebration in 1963, for the first time in centuries, the volcano erupted, killing thousands of people and sending lava coursing over eastern Bali. Miraculously, Besakih Temple remained untouched.

biblioGraPHY

Bigger Secrets by William Poundstone (New York: Houghton Mifflin, 1986)

Biggest Secrets by William Poundstone (New York: Quill, 1993)

"Bizarre Stuff You Can Make in Your Kitchen" by Brian Carusella, http://freeweb.pdq.net/headstrong/default.htm

The Book of Lists by David Wallechinsky, Irving Wallace, and Amy Wallace (New York: Bantam, 1977)

Can It Really Rain Frogs? by Spencer Christian (New York: John Wiley & Sons, 1997)

Chariots of the Gods? by Erich Von Däniken (New York: Bantam, 1970)

The Concise Oxford Dictionary of Proverbs by John Simpson (Oxford, England: Oxford University Press, 1992)

Dictionary of Trade Name Origins by Adrian Room (London: Routledge & Kegan Paul, 1982)

The Dorling Kindersley Science Encyclopedia (New York: Dorling Kindersley, 1994)

Eyewitness Books: Crystal and Gem by R. F. Symes and R. R. Harding (New York: Knopf, 1991)

Famous American Trademarks by Arnold B. Barach (Washington, D.C.: Public Affairs Press, 1971)

The Guinness Book of Records edited by Peter Matthews (New York: Bantam Books, 1998)

"Have a Problem? Chances Are Vinegar Can Help Solve It" by Caleb Solomon (*Wall Street Journal*, September 30, 1992)

How in the World? by the editors of *Reader's Digest* (Pleasantville, New York: Reader's Digest, 1990)

How to Play with Your Food by Penn Jillette and Teller (New York: Villard, 1992)

How to Spit Nickels by Jack Mingo (New York: Contemporary Books, 1993)

Janice VanCleave's 200 Gooey, Slippery, Slimy, Weird & Fun Experiments by Janice VanCleave (New York: John Wiley & Sons, 1993)

The Joy of Cooking by Irma S. Rombauer and Marion Rombauer Becker (New York: Bobbs-Merrill, 1975)

Jr. Boom Academy by B. K. Hixson and M. S. Kralik (Salt Lake City, Utah: Wild Goose, 1992)

Marbling by Diane Vogel Maurer with Paul Maurer (New York: Friedman/Fairfax, 1994)

Marbling Techniques by Wendy Addison Medeiros (New York: Watson-Guptill, 1994)

Martin Gardner's Science Tricks by Martin Gardner (New York: Sterling, 1998)

100 Make-It-Yourself Science Fair Projects by Glen Vecchione (New York: Sterling, 1995)

Panati's Extraordinary Origins of Everyday Things by Charles Panati (New York: Harper & Row, 1987)

Paper Art by Diane Maurer-Mathison with Jennifer Philippoff (New York: Watson-Guptill, 1997)

Paper Craft School by Clive Stevens (Pleasantville, N.Y.: Reader's Digest, 1996)

Physics for Kids by Robert W. Wood (Blue Ridge Summit, Penn.: Tab Books, 1990)

The Safe Shopper's Bible by David Steinman and Samuel S. Epstein, M.D. (New York: Macmillan, 1995)

Science Fair Survival Techniques (Salt Lake City, Utah: Wild Goose, 1997)

Science Projects about Light by Robert Gardner (Hillside, N.J.: Enslow, 1994)

Science Wizardry for Kids by Margaret Kenda and Phyllis S. Williams (Hauppauge, N.Y.: Barron's, 1992)

Shake, Rattle and Roll by Spencer Christian (New York: John Wiley & Sons, 1997)

South-East Asia Handbook by Stefan Loose and Renate Ramb (Berlin, Germany: Stefan Loose Travel Books, 1983)

Structure of Matter by the editors of Time Life (New York: Time Life, 1992)

365 Simple Science Experiments by E. Richard Churchill, Louis V. Loeschnig, and Muriel Mandell (New York: Black Dog & Leventhal, 1997)

What Makes the Grand Canyon Grand? by Spencer Christian (New York: John Wiley & Sons, 1998)

Why Did They Name It . . . ? by Hannah Campbell (New York: Fleet, 1964)

acknowledgments

I am grateful to Jennifer Repo, Erin Stryker, Dolores McMullen, and John Duff at Perigee Books for their enthusiasm, passion and excitement for this book. They all made working on this project a sheer joy. I am also indebted to Amy Schneider for her astounding copyediting expertise.

A very special thanks to Jeremy Solomon at First Books for sharing my unbridled exuberance for this book and for his inspiring professionalism and zealous perseverance.

I am also grateful to my father, Bob Green, for helping me build a volcano in the fourth grade, Zen poet Jeremy Wolff for perfecting the hydrogen experiment, Tony Salzman for sharing his firsthand experiences with steel wool and D batteries, and Jim Parish for suggesting the idea in the first place.

Above all, all my love to my wife, Debbie, and my daughters, Ashley and Julia, for letting me make a huge mess in the garage.

about the Author

Joey Green, author of *Polish Your Furniture with Panty Hose*, *Paint Your House with Powdered Milk*, and *Wash Your Hair with Whipped Cream*, got Jay Leno to shave with Jif peanut butter on *The Tonight Show*, Rosie O'Donnell to mousse her hair with Jell-O on *The Rosie O'Donnell Show*, and had Katie Couric drop her diamond engagement ring in a glass of Efferdent on *Today*. He has been seen polishing furniture with Spam on *Dateline NBC*, cleaning a toilet with Coca-Cola in *The New York Times*, and washing his hair with Reddi Wip in *People*. Green, a former contributing editor to *National Lampoon* and a former advertising copywriter at J. Walter Thompson, is the author of more than a dozen books, including *The Zen of Oz: Ten Spiritual Lessons from Over the Rainbow*, *Selling Out: If Famous Authors Wrote Advertising*, and *The Bubble Wrap Book*. A native of Miami, Florida, and a graduate of Cornell University, he wrote television commercials for Burger King and Walt Disney World, and won a Clio Award for a print ad he created for Eastman Kodak. He backpacked around the world for two years on his honeymoon, and lives in Los Angeles with his wife, Debbie, and their two daughters, Ashley and Julia.

Visit Joey Green on the Internet at
www.wackyuses.com